AF722557

The Mystery of Alpha

Copyright © 2025 by S.N.W. Hilt.
All rights reserved.

ISBN 979-8-89487-012-0

The Mystery of Alpha

by

S.N.W. Hilt

Copyright © 2025 by S.N.W. Hilt.
All rights reserved.

ISBN 979-8-89487-012-0

Golden Tao Publishing
Angel Fire, NM
USA

Dedication

This book is dedicated to my family that helped on this lengthy road of discovery: Cindy, Nathaniel, Zachary, Sybil, Eric, Joshua, Teresa, Steffen, Hannah, Anders, Angelo, John and Susan.

Contents

Preface xi

1. Introduction 1

2. The Mystery of Alpha, $\alpha \cong \frac{1}{137}$, from physics 7
    2.1 The mystery of Alpha in Physics 7
    2.2 Fractal Reality 9
    2.3 The Maximum Information Emanation Hypothesis 11
    2.4 Unit-norm propagation 13
    2.5 Finding Alpha (α) in Mathematics of Unit-norm Emanation ... 15
        2.5.1 Derivation of $\{\alpha\}$ 15
        2.5.2 Derivation of $\{\alpha, \pi\}$ 16
        2.5.3 Derivation of $\{\alpha, \pi, C_\infty\}$ 17
            2.5.3.1 Maximum information at the edge of chaos 20
            2.5.3.2 Fractal Echo Phenomenon -- kabbalah 21
    2.6 The Fundamental, Chiral, Emanation Process 21
    2.7 The Full (Achiral) Emanation Process 22
        2.7.1 Universal Thermality 22
        2.7.2 Achiral Emanation types and Tarot Deck structure 22
    2.8 The Standard Model, Standard Cosmology, Implications 23
    2.9 The 'Alpha numbers' (further details in App. A) 24

3. Multitudinous Fractals Incarnate 25
    3.1 Alpha from stable, high complexity, long-term cycles 25
        3.1.1 The Earth-Sun magnetosphere cycle (22 Earth cycles) 26
        3.1.2 Other observations of systems with 22 'states' 27
        3.1.3 Fractal G, Scale Invariance, Renormalization Group 28
    3.2 System properties from long-range correlations 30
        3.2.1 The Ising Model 31
        3.2.2 Scaling in Large Language Models 44

4. The Mystery of Life (a biochemical encoding description) 47
    4.1 Information encoding molecules, autocycles, and PAH World . 47
    4.2 The original cellular encoding: RNA World 56
    4.3 The modern encoding: 22 AA's and 137-exon gematria 61

4.4 The Information and Metabolic Cycles (details in App. B) 69

5. The Development of Intelligent Life (neural encoding description). 79
5.1 Statistical Learning and Evolving Complexity 80
5.2 Optimal Learning on a Neuro-Manifold 80
5.2.1 Differential Topology—Natural Manifold 80
5.2.2 Differential Geometry –Natural Geometric Structures ... 85
5.3 Amari's Dually Flat Formulation 88
5.3.1 Generalization of Pythagorean Theorem 91
5.3.2 Projection Theorem, Divergence and Link Formalism ... 92
5.4 Neuromanifolds 93
5.5 LLM Learning 96
5.5.1 Learning with long-range correlations 96
5.5.2 The self-attention mechanism 97
5.5.3 Learning with scaling behavior and knowledge capture .. 99
5.5.4 The Reticular Activating System (RAS) 101

6. The Evolution of Human Communication (symbolic encoding) 103
6.1 Lateralization 104
6.2 Timeline of Hominin Tool and Language Developments 108
6.3 Mixed hominin communities 111
6.3.1 Evidence of coexistence 111
6.3.2 No Y-chromosome is from Neanderthal lineage 113
6.3.3 Technology Transfer (excluding language) 114
6.3.4 Proto-Linguistic Hominin Developments 115
6.4 Invention and early use of music in hominins 116
6.4.1 Early Evidence of Music 116
6.4.2 Theories on the Origins of Music 117
6.4.3 Alpha music: 22-states of tonality,12TET, 24TET 117

7. The Mystery of Divination: Kabbalah and Tarot 119
7.1 The Universality of Alpha and the Sacred Numbers 119
7.2 The Gematria Divination Game 119
7.2.1 Assyrian Tree-of-Life: {4, 9, 22, 29, 32, 56, 78, 137} 122
7.2.2 Hebrew Tree of Life: {4, 10, 22, 32, 56, 78, 137} 127
7.3 Divination methods indicated 132
7.4 Chronology of ancient forms and practices of divination 132
7.4.1 Interpretive Divination (no Alpha, no refinement) 132
7.4.1.1 Augury 132
7.4.1.2 Geomancy and Bone (drop) Divination 132
7.4.1.3 Scrying 132

7.4.1.4 Oracle Bones 132
7.4.1.5 Dream Interpretation.. 133
7.4.2 First Wave: Yarrow Sticks 133
7.4.3 Second Wave: Livers 134
7.4.4 Third Wave: I-Ching Hexagrams 134
7.4.5 Fourth Wave: use of Dice.. 134
7.4.5.1 Runes.. 135
7.4.5.2 The Game of Ur 135
7.4.5.3 Divination with Hebrew Tree of Life Diagram 136
7.4.6 Fifth Wave: use of sheet dice (Tarot)............................... 137
7.4.7 Sixth Wave (from Babylon): use of Zodiac 138
7.4.8 Seventh Wave (from Ottoman/Istanbul): Zodiac 138
7.4.9 Eighth Wave: Emergence of Tarot in the West................. 138
7.4.10 Ninth Wave: Tarot and Kabbalah 139

8. The Mystery of the Logos 141
8.1 The Classical Logos (Hellenic, Jewish, Christian, Islamic)...... 142
8.2 Text Analytics identification of the Logos 143
8.2.1 Machiavelli's Polysemy on *Fortuna* and *Virtu* 143
8.2.2 Coleridge's Hidden Polysemy on *Logos*......................... 146
8.2.3 Shakespearean Phrases 148
8.3 The Logos: From proto-civilization to Modern Era................. 149
8.3.1 The Blue Sea Proto-Civilization Hypothesis 149
8.3.2 The proto-civilization Göbekli Tepe................................ 151
8.3.3 Civilization from Neolithic to Modern Era...................... 152
8.3.4 Early Civilization, Law, and Religion 154
8.4 The Logos: 'word-centric' reality 155

9. Emanation and Emergence of Consciousness 159
9.1 Language, Narrative, and Consciousness.............................. 159
9.2 Tarot-complete Narrative Generation.................................... 160
9.3 Narrative with Data: The Book of Enoch................................ 161
9.3.1 Repeated story elements and refinements 162
9.3.2 Externalized collective unconscious................................ 162
9.4 (Internalized) Collective Unconscious (Jung) 163
9.5 Sequential Nature of Consciousness 163
9.5.1 Physics Perspective 166
9.5.2 Neuroscience Perspective 167
9.5.3 Philosophical Perspective 167
9.6 Awakening Consciousness.. 168
9.6.1 Sartre (Individual Logos).. 168

9.6.2 Eisley (Civilizational Logos) 168

10. Eucatastrophe and the Advent of Human Spirituality 171
10.1 Life is a kabbalistic optimum 171
10.2 Man is in the image of God (alpha evolution)........................ 171
10.3 CS Lewis, Chomsky – there is inherent human structure 172
10.4 CS Lewis provides an early Multiverse argument.................. 172
10.5 Emanation has a Unitarian/Trinitarian structure..................... 173
10.6 Emanation and Harmony (Interconnectedness)...................... 173
10.7 Fractal emanation and transcendent consciousness 173
10.8 Maximal emanation narrative and the soul............................ 173
10.9 Emanation links Science and Religion – Renaissance 2.0 173

Appendix 175
A. Emanator Theory Overview 177
B. Biology Information Processing 213
C. Narrative world: multiverses, magic, monadic travel 255

References 257

Preface

The mystery of alpha occurs in both physics and in gematria-based divination methods (word scoring systems). In Physics alpha is also known as the fine structure constant and is slightly less than 1/137. This is fundamental constant in physics, like the speed of light, but it is even more fundamental in that it is dimensionless (i.e., its just a number). In ancient gematria-based divination methods, on the other hand, the score 137 occurs for special words. The mystery of alpha in both of these instances is that the origin of these two special numbers, or one (if the inverse is counted as revealing only one special number) is unknown. In the physics, however, a description based on maximal information emanation (Emanator Theory) has identified the origins of alpha (and its derivation). Three independent derivations of alpha result from the Emanator Theory description, all in agreement with experimental observation to the highest current level of experimental observational precision (over a dozen decimal places). An overview of Emanator theory is given in the appendix and a non-math overview is given in the first chapter after the introduction. In explaining the mysterious alpha in physics, Emanator Theory also indicates a collection of fundamental structural numbers in the theory, that enumerate the different generative 'motions' in the theory. Those numbers, referred to as the 'Alpha Numbers' are: {4, 9 or 10, 22, 29, 32, 56, 78, ~1/137}. As will be seen, these are also the 'Sacred numbers' or 'Kabbalah numbers' from ancient divination methods, where precisely this set of numbers is encoded in the Assyrian Tree of Life (as will be shown), and in the Kabbalistic Tree of Life. This set of numbers also occurs (is encoded) in the familiar Tarot Deck (itself derived from the aforementioned divination systems and inheriting the Sacred numbers in this process), e.g., the Tarot deck has 78 cards, 4 suits, 22 major arcana, 56 minor, etc. The mystery of alpha in physics, seemingly solved, is now compounded by this amazing correspondence between Alpha numbers and the Sacred Numbers (and the mystery of the Kabbalah). This larger mystery will be solved in the book that follows, starting from the physics description (where the sacred numbers are clearly delineated) and then going to the development of intelligent life, then to the development of linguistic capabilities, up to the ability to play divination 'games'. What is revealed in this exploration is not only the explanation of alpha and the sacred numbers, but also an understanding of the mystery of the Logos, the emergence of human consciousness, and a linkage thereby between science and philosophy/religion. This linkage suggests we are on the cusp of a new Renaissance, where science and religion, philosophy, art, and humanities are about to have explosive growth in a new era of synergistic exploration.

Chapter 1. Introduction

This book describes how the mystery of Alpha begins to unravel with application of the maximum information emanation (MIE) hypothesis [1-10], where Alpha, known as the fine structure constant in physics, is obtained both theoretically and computationally (in a limit process), and where this value of Alpha is in precise agreement with the physically observed value of Alpha. In the process of obtaining the theoretical derivation for Alpha a collection of structural parameters are also indicated. What is remarkable is that there is not just the correspondence of 1/Alpha being approximately 137, a Kabbalistic number, but that the entire set of Alpha numbers corresponds with the set of Kabbalistic numbers. The mystery of Alpha from physics, seemingly almost solved in the mathematical context of MIE (or "Emanator Theory" [1]) has now grown more complicated, including, to some extent, the mystery of the Kabbalah as well. How can modern theoretical physics intersect with ancient gematria divination practices, still practiced in Kabbalah today, but themselves traceable to the gematria systems used by the ancient Babylonians, Assyrians, and Sumerians? This book will not only explain this connection but will describe how radically different domains, such as science (physics, biology), mathematics, language, and divination 'games', can all have common structural elements when optimized in various scenarios. This remarkable interconnectedness or universality is precisely how Coleridge describes the Logos [11]. Having run the gamut of topics from Alpha to the Kabbalah to the Logos, I'd be remiss if I didn't continue the run and touch on ramifications for a new understanding of consciousness and for the advent of human spirituality (Chapter 10). If this sounds like an "Immense Journey", delving into archaic beginnings and connecting to modern day, and beyond, reminiscent of Loren Eisley's book of that title [12], then that's because it is (and discussion of Eisley's book is given later as well). So, gird thyself for an amazing ride (or, to be more formal: "Stand therefore, having your loins girt about with truth, and having on the breastplate of righteousness." Ephesians 6:14, KJV). To begin, let's start with when the mystery was first (re)discovered.

The mystery of Alpha is encountered in two domains: physics, since spectral observations in the 1890's, and in gematria word divinations practiced by a number of groups as far back as 2500 BC (referenced in the Weld-Blundell Prism from Sumer [13], 2100BC, and in the Babylonian and Akkadian "Epic of Gilgamesh" tablets dating to 2100 BC also). In physics the mysterious constant alpha has the value approximately 1/137, while in Kabbalistic gematria word scoring, words that score 137 are significant, such as the word Kabbalah itself, e.g., the name of the Jewish mystical practice itself (and the only known meaning of that word is as that name). There is meaning in the root word "Kabbal", however, which means "to receive", where "Emanation" is often described as a process of receiving in both Kabbalistic mystical descriptions and the Emanator theory mathematical-physics description.

At first introduction of the constant Alpha = ~1/137 in the physics community it was noted that this was tantalizingly close to the Kabbalistic number of 137. This did not clarify matters, however, as there were now two numbers that were mysteriously appearing that appeared to be the same number just given in the inverse in the other domain. The unknown origin of either number and their possible relation had only deepened the mystery of Alpha.

In this book we start with a non-expert description of the results from "Emanator theory" from theoretical physics, where simple application of the 'maximal information emanation' (MIE) hypothesis leads to Standard Model Physics, as currently known, but WITH an explanation (derivation) for the value of Alpha. The theoretical prediction of Alpha from this agrees with that observed experimentally to incredibly high precision (at least 15 significant figures). In the description of Emanator Theory we will find not only the appearance of the 'coupling constant' Alpha, as indicated, but also structural elements in the system's propagation that involve the numbers {4, 9, 22, 29, 32, 56, 78}. This collection of numbers will sometimes be referred to as the 'Alpha numbers'. Details on emanator theory are available in [1-10], but for convenience the key derivations are also given in Appendix A. The discussion of emanator theory provided in the next chapter, Chapter 2, is then at a high level (no math), focusing on the structural properties indicated, according to the numbers given, as well as describing the fractal nature of reality indicated by the theory.

As mentioned, there is the parallel mystery of ~1/Alpha=137 appearing in gematria word divination since antiquity. If we turn to the Hebrew gematria variant used by the Kabbalists (Jewish mystics), we find that

137 was significant, as well as the numbers {4, 10, 22, 32, 56, 78}. In Chapter 7 we will see that the Kabbalah Tree of Life diagram encodes these numbers in a straightforward way. Comparison with the numbers above shows remarkable agreement between the MIE selected physics 'numerology' number-set based on the 'Alpha' numbers and that of the 'Kabbalah numbers'. This is an even more shocking correspondence than that between just ~1/137 and 137 first seen in the 1890's. It is hard to do better than this remarkable correspondence, but it can be done, and exactly. Chapter 7 also shows the Tree of Life diagram used by the ancient Assyrians (who also practiced gematria). The tree symbol seems very elaborate and very specific in its elements, and this is for good reason, as it precisely encodes all of the Alpha numbers, including the number '29' represented in the number of flowers on the tree.

Clearly the physics-based numerology encompassed by {4, 9, 22, 29, 32, 56, 78, ~1/137}, which is mathematically proven to result in a maximum information emanation (MIE) process, can also be found to result in other processes that are optimized for maximal information transmission. What we are seeing with gematria is the end result of a word scoring game that evolved over time to arrive at a form that generates distinctive scoring. On that basis, it is a communication channel optimized over time (with scoring variations in the divination 'game'), with the familiar 137 resulting. Aside from the gematria game, the Tree of Life encoded properties of the world (in either Assyrian or Hebrew formulation) with the other number encodings indicated. So the bizarre correspondence between the mathematically derived Emanator theory, which is exact, with the numerology encoded by the ancients in the Tree of Life diagrams (and related interpretations) is profound. How did the ancients uncover the same numerological structure that undergirds Emanator theory? In both cases we see that the Alpha constants describe a universal solution, of optimized (maximal) information transmission, or emanation, or structure realization in general. So perhaps the Alpha constants are emergent in any scenario where there is long-term information transmission, that steadily optimizes over time. This would be the case for spoken language as well as written. This would be the case for long-term cycles (such as the Sun-Earth 22 year magnetosphere cycle). This would appear in nature's encodings for proteins (with 22 amino acid building blocks or 'letters').

So, the answer to the mystery of Alpha, from both modern and ancient encounters, appears to be that the Alpha constants are emergent in any scenario where there is long-term maximal information transmission. For this to be true however, at the level of human written communication, overlaid with the gematria game, then it is probably true (as an emergent construct) in more direct scenarios, such as spoken language itself, or possibly intelligence before that. In essence, the notion that reality is a long-term emanation of maximal information, that gives physics as we know it, has numerous 'down-stream' effects. The same mechanism for emergent structures (with Alpha constants) can now be replayed in every instance of long-term propagation, including life as a long-term process, of the emergence of intelligence, of communication, and possibly the nature of consciousness itself.

The mystery of alpha as seen in physics is resolved in Chapter 2, but to get to addressing this mystery in the context of Kabbalah in Chapter 7, there is much intervening territory to explore with an eye towards Alpha numbers, the ramifications of fractal properties that are also indicated, and establishing the framework of life that would eventually develop communication at the level performed by humans in Chapter 7. To this end, Chapter 3 will focus on fractal reality with long-term cycles or long-range correlations. In this Chapter we will see the impact of fractal, scale-invariant, physics in terms of renormalization group properties critical to making quantum field theory well-defined. These same tools are used in identifying global structure emergence and phase transitions in statistical physics; and likewise identifying long-range correlations in statistical learning, such as with the breakthroughs in LLMs in recent years [14,15].

One of the oldest long-term processes is life itself. In Chapter 4 is an overview of Life from abiogenesis to current day. Again, there are surprising occurrences of the alpha numbers now that they are known and can be easily spotted (such as the number 22 occurring in the 22 amino acids that code for all proteins, the 'words' of life, compared with the 22-letters in Hebrew, and all ancient scripts, used to make actual 'words'). Chapter 5 describes Intelligent Life as a natural outcome of Life itself, with attention to how 'learning' might occur with a neural net. Chapter 6 focuses on human communication as a long-term process and how the alpha numbers might be emergent in this context. There are numerous other modes of communication than language as well, most notably divination methods and music.

As mentioned, the mystery of alpha is also the mystery of the Kabbalah, and in Chapter 7 we return to the discussion of gematria and the history of divination methods in general. Of the divination methods we see that the Tarot will stand out as a direct mapping of the Alpha numbers, derived from the matching Kabbalistic numbers, probably developed in Babylon. The Tarot deck has 78 cards split into major arcana (22) and Minor Arcana (56) the latter split over 4 suits. Thus, many of the alpha numbers are directly represented. Does Tarot match with the alpha numbers, themselves part of a 'complete' physical description for evolution? We will see that emanation involves 78 'generators' of evolution (but 22 of these 'generators' make no change in their motion, so effectively describe constants or fixed set of states in the system), such that the Tarot deck offers the possibility of 'complete' narrative generation from its set of cards. This is discussed in Chapter 8 in an effort to describe a 'stream of consciousness' via a sequence of mental states that correspond with the Tarot cards.

The evolution of physical reality, both cosmic and biological, eventually followed by evolution of our hominin forbears, has resulted in our ability to speak, have language, and even play gematria 'games'. With emergence of language there is often associated the emergence of consciousness itself (Leibnitz has much to say on this as will be seen). So, in Chapter 9 is tackled the notion of consciousness from the perspective of language and how that might define our consciousness ('Alpha consciousness' to be described). Mention of hominin forbears is not made lightly, either, as early acquisition of language and divination games (and burial practices, etc.) may have been adopted from our more ancient hominin forbears. The brain asymmetry (lateralization) driving human language capabilities may have been further accelerated via selection on children of hominins resulting from interbreeding (such as Homo Sapiens and Neanderthal or Denisovan). Chapter 9 is, thus, an attempt to explain the advent of human consciousness in the context of the emergent Alpha communication properties indicated.

A unique perspective on consciousness is thus obtained in solving the mystery of Alpha, and seeing how this impacts human communication (e.g., in both external and internal dialogues). Just as a scientific/mathematical bridge to metaphysical notions of consciousness has been made possible by the understanding of maximal information emanation and the emergent Alpha numbers/structures described, so too can this bridge be extended even further to describing the advent of human spirituality itself. This is done in Chapter 10 in a restricted sense as will be made clear. This extension of emergent Alpha consciousness to the topic of spirituality is not too much of a stretch since human religiosity/divination has evidently been hardwired into the human brain as far back as the rudiments of language itself. What is remarkable is the familiarity of the variety of reasonable outcomes indicated, in the context of modern physics with its Many Worlds models. These outcomes and indications for what might be human 'spirituality' will be very familiar, they are the central tenets of the worlds major religions, philosophies, and psychologies.

Chapter 2. The Mystery of Alpha, $\alpha \cong \frac{1}{137}$ from physics

Emanation Physics [1-10] provides an explanation for the origin of the mysterious physical parameter alpha ($\boldsymbol{\alpha}$), a.k.a., the fine structure constant, where $\boldsymbol{\alpha^{-1} = 137.05999}$ ... Other fundamental constants of the theory are indicated, including their enumeration and groupings. A synopsis is given here, with a more mathematically detailed overview given in Appendix A. We will see that there are 22 fundamental constants and there are 78=22+56 elements, or generators, in a complete emanation process. The 56 describes 4 chiralities, where 56=4x14. Emanation is in a 10 dim subspace of a 32 dim space, where perturbation that retains chirality entails a perturbative chiral emanation that extends into a 29 dim subspace of the 32 dim space. The emanation process is unit norm, so a 9-dim subspace within the 10-dim space when this is enforced (without perturbation). Of the 78 generators, a subset splitting of 8+56=64 also exists with properties. From this we get what will be referred to as the 'Alpha numbers': {4, (9 or 10), 22, 29, 32, 56, 78, 137, and alpha}, which will be shown to be identical with the 'kabbalistic numbers' later.

2.1 The mystery of Alpha in Physics

In the 1880's Balmer and Rydberg find that spectral lines obey a simple formula, with a single critical constant, universally occurring, now known as the Rydberg constant ($R_H = 109721.6\ cm^{-1}$). This was an astonishing result at the time [16]. Sommerfeld noted the almost cabbalistic underpinnings of the mathematics (in his book Atombau and Spektrallinien [17], Sommerfeld referred to the Rydberg top square equation as a 'cabbalistic' formula). Wolfgang Pauli, a student of Sommerfeld's, shared his keen interest in the origins of α and turned it into a life-long obsession. So much so, that it practically drove him mad, to where he sought the help of famed psychoanalyst Carl Jung, with whom he eventually partnered to try to solve the mystery of α (the madness is contagious). From Pauli's Nobel Prize Lecture [18]:

> "From the view of logic my report on 'Exclusion principle and quantum mechanics' has no conclusion. I believe it will only be possible to write the conclusion if *a theory will be established which will determine the value of the fine structure constant* and will thus explain the atomistic of electric fields actually occurring in nature." (emphasis mine)

The obsession with α continued with the next generation of great Physicists as well, particularly Feynman, who said [19]:

> "There is a most profound and beautiful question associated with the observed coupling constant, *e* – the amplitude for a real electron to emit or absorb a real photon. It is a simple number that has been experimentally determined to be close to 0.08542455. (My physicist friends won't recognize this number, because they like to remember it as the inverse of its square: about 137.03597 with about an uncertainty of about 2 in the last decimal place. It has been a mystery ever since it was discovered more than fifty years ago, and all good theoretical physicists put this number up on their wall and worry about it.) Immediately you would like to know where this number for a coupling comes from: is it related to pi or perhaps to the base of natural logarithms? Nobody knows. It's one of the greatest damn mysteries of physics: a magic number that comes to us with no understanding by man. You might say the "hand of God" wrote that number, and "we don't know how He pushed his pencil." We know what kind of a dance to do experimentally to measure this number very accurately, but we don't know what kind of dance to do on the computer to make this number come out, without putting it in secretly!"

Alpha is a critical constant in modern science, acting as a key parameter in the most important physics equations. Alpha (in the form 1/Alpha≈137) is also the key parameter in Kabbalah and other ancient numerological systems (Tarot, Qabalah). This would just be an interesting coincidence if it didn't extend to a set of 'mystical' numbers (not just 137). Let's briefly elaborate on the set of numbers. They are: {4, 10, 22, 29, 32, 56, 78, 137}. In physics, we have 4D spacetime; we have 10D emanation; we have 22 parameters in the extended Standard model; 29 is the dimension of doubly chiral unitary propagation in the larger 32D space; 56 and 78 are the number of generators of motion in emanator theory, and 137 is 1/Alpha, the most mysterious physics constant of all. In Kabbalah, we have 4 Worlds; 10 sefirot; 22 letters; and various ways to numerologically describe/include the numbers 32, 56, 78, and 137. Interestingly, the number 29 is missing from the (Hebrew) Kabbalah, but is indicated clearly in the Assyrian Kabbalah. The delineation of the set of sacred numbers has been made most clear from discoveries in modern physics, so what follows starts with an overview of the physics history of discovery as regards Alpha and the other 'sacred' numbers. The same review will then be done for the Kabbalistic numbers (Chapter 7), but we don't have a beginning for the Kabbalah history, nor a clear delineation of the sacred numbers. We will find, however, that the set of all the main kabbalistic numbers will be the same as the physics set.

2.2 Fractal reality

Fractal reality refers to the idea that certain patterns and structures in nature exhibit self-similarity across different scales, a concept derived from fractal geometry. This notion has implications in various fields, including physics, biology, and cosmology. Fractal geometry studies shapes and patterns that are complex and self-similar. Fractals can be found in natural phenomena such as coastlines, mountain ranges, clouds, and snowflakes, where similar patterns recur at different scales [20-24].

In fractal structures, a small portion of the pattern resembles the whole, i.e., there is self-similarity. This property can be observed in various biological systems, such as the branching of trees and the structure of blood vessels [25-28].

When there is unlimited self-similarity there is formal, mathematical, scale invariance, for which amazing properties are proven to exist.. Fractals often exhibit scale invariance, , so, for example, their statistical properties remain consistent regardless of the scale at which they are observed [29-33]. This property is highly relevant in physics, particularly in the study of critical phenomena [34], phase transitions, statistical field theory, and quantum field theory [35].

Fractal concepts have been applied to understand complex systems, such as turbulence, the distribution of galaxies, and the behavior of materials at different scales. Note that the distribution of matter in the universe can exhibit fractal-like patterns [36]. Some theories in cosmology suggest that the universe may have a fractal structure on large scales, with galaxies and clusters of galaxies forming a web-like pattern [37,38]. This challenges traditional views of uniformity in the cosmos [39,40].

The idea of fractal reality raises new questions about the nature of reality itself, suggesting that the universe may be more interconnected and complex than previously thought, as indicated in [1].

In Emanator theory a complex trigintaduonion is considered, an object that is similar to $\mathbb{C}^{32}$, where each $\mathbb{C}$ in the 32-factor product space participates in a mapping $\mathbb{C} \to \mathbb{C}$, with limit an iterative mapping with (proven [41]) universal limits. These same 'universal' limits are exhibited with the $\mathbb{C} \to \mathbb{C}$ description that defines the Mandelbrot Set [20]. The Mandelbrot Set is more than just a familiar point of reference for many readers, it is representative of an actual solution relevant to MIE (as a local isomorphism in the critical emanation mapping that must be 'maximal'). This is because the fractal boundary of the Mandelbrot Set, theoretically between 1 and 2 dimensions, is maximal at 2 dimensions [42].

To see the fractal connection to Emanator theory, consider the Mandelbrot iterative mapping based on the function $z_n = (z_{n-1})^2 + c$. For choice c and initial $z_0=0$, if $z_\infty \rightarrow \infty$, then that c is outside the set, otherwise, if remains bounded, then it's in the (Mandelbrot) set [20]. This is an example famous for its beautiful fractal images and mathematical properties. The largest c value (at the edge of chaos) is known as the bifurcation parameter and is $c_\infty = 1.401155189....$ The maximum allowable 'perturbation' for z (not z^2) would then be $(c_\infty)^{(1/2)}$. In the trigintaduonion propagation we discover in what follows we have chiral propagation in the 32 dim trigintaduonion space, where the real dimension is fixed by the unit-norm property, leaving 29 'free' imaginary dimension/parameters, since two more are selected for a specific chirality. If we allow the same maximal bifurcation parameter as a factor for each of the 29 free dimensions (and for a small imaginary part overall), a precise relation will be obtained according to the exact form of the trigintaduonion emanation. This provides a third relation that provides a derivation for Alpha.

There are, thus, three Alpha relations in total:

(i) A computational limit relation $\{\alpha\}$ for α alone due to the limit on chiral trigintaduonions with maximum perturbation;

(ii) A relation $\{\alpha, \pi\}$ due to maximum perturbation occurring when noise has maximum antiphase;

(iii) The $\{\alpha, \pi, c_\infty\}$ relation due to maximum information flow occurring at the edge of chaos.

To recap -- the trigintaduonion reformulation from emanator theory provides a deeper mathematical foundation to describe the underlying information 'propagation' (with initial propagation being referred to as 'emanation'), with a precise derivation, and with a unit-norm perturbative limit that leads to an iterative-map-like computed α (a limit that is precisely related to the Feigenbaum bifurcation constant and thus fractal). The familiar Mandelbrot set: $f(z) = z^2 + c$ (complex) has C_∞as a limit value on c in the iterative map for stability. Similarly, and directly relevant here, for the real one-parameter map $f(x) = a - x^2$, C_∞ is the (universal) limit value on a [41]. Since repeated chiral emanation steps can be described as an iterative mapping, with unit-norm constraint resulting in a quadratic relation on components, we expect the Feigenbaum universal bifurcation parameter, C_∞, to appear according to the number of independent dimension in a chiral trigintaduonion emanation step and the precise form of the "emanator" construction. The number of effective dimensions is shown to be 29 (plus a little more from perturbation analysis), where a relation between α, π and C_∞results that is dependent on the choice of (achiral) emanator. The emanator identified by the $\{\alpha, \pi\ C_\infty\}$ relation is used to explore random walks in the Trigintaduonion space during emanation (shown to be Martingale empirically) and explore noise additivity effects.

2.3 The Maximum Information Emanation (MIE) Hypothesis

Emanator theory stems from the Maximum Information Emanation (MIE) Hypothesis. The definition of information is context dependent, so how the MIE hypothesis will manifest depends on circumstance. We start with the fundamental notion of the quantum propagator, for which mathematical 'propagators' satisfy unitarity. We seek to extend this foundational element so start by asking what is the highest Cayley algebra that can remain unitary – where the answer comes down to the highest order division algebra, which are the octonions. What if we change the desired property from unitary to unit-norm preserving? Then we can extend 2 more dimensions beyond the 8D octonion algebra to a chiral subspace of the trigintaduonions that is 10D [2].

The 10D chiral trigintaduonions are then identified as the maximal information 'carriers' or emanators, operationally like the quantum propagators in a larger theory, where evolution of the system will be shown to result from sums on paths of emanators (similar to quantum evolution in terms of a path integral on propagators 'steps'). Having posited this maximal construct we see the classic signs of "asking the right question" since we get a variety of clear results:

(1) the chiral emanator is manifestly Lorentz Invariant (when using BiSedenion form).
(2) chiral emanation involves a 10D element in a 32D space (trigintaduonions) for which maximum perturbation (still permitting unit-norm transmission) is by the amount 'alpha' – we therefore have the mysterious alpha by a computational definition.
(3) chiral emanation with perturbation does not have effective dimension 32D due to chiral and other constraints -- noise budget analysis or

(equivalently) Kato Rellich operator analysis, both indicate effective dimension slightly greater than 29 referred to as "29*".
(4) the chiral emanator indicates non-perturbative 'motion' in a 10D subspace of 32D, suggesting 22 constants of the motion. This is revealed to be true in the mathematical formalism as there are 22 types of emanation that result in no change to the base trigintaduonion describing the system. This, in turn, suggest that the emanator theory will have 22 parameters.
(5) By using the split form of the trigintaduonions, we not only have manifest Lorentz invariance, we also have an exact algebraic split to a space that is simply the direct product of 29* real dimensions (not a local approximation to such). This is important because the fundamental existence of a complex structure (two such layers of structure will be indicated) means that we trivially have the extension $\mathbb{R}^{29^*} \rightarrow \mathbb{C}^{29^*}$. Suppose we have point-like singular elements in $\mathbb{C}^{29^*}$, such will occur due to zero-divisors (zd's) in the 32D trigintaduonion space. To achieve a maximal domain of analyticity (an application of MIE), we must remove the zd-singularities. In doing so we obtain point-like matter in the theory and a small-h constant that enters the sum on emanator paths just as Planck's constant in the sum on propagator paths in the quantum formulation – suggesting that these small-h numbers are related.
(6) Three derivations of alpha are obtained: (i) $\{\alpha\}$ based on the maximal perturbation for which chiral emanation retains the unit-norm property; (ii) $\{\alpha, \pi\}$ based on the maximal noise transmission on a chiral emanation path; and (iii) $\{\alpha, \pi, C_\infty\}$ based on the maximal noise transmission on an achiral emanation path (where maximal emanation is at "the edge of chaos" which is defined according to Feigenbaum Universality [41]).
(7) At component level in the emanation product, using 100's of millions of computational steps, we see an excellent asymptotic fit to random walk behavior. Since a random walk is a Martingale process, this strongly suggests that the achiral emanation process is Martingale. In turn, the projected quantum process (standard theory) would retain the imprint of that Martingale process.
(8) The achiral emanator, a sum of achiral emanation paths, can be shown to have the mathematical form $\sum \exp(i\mathbb{H} \times \mathbb{O}) \rightarrow \mathbb{C} \times \mathbb{H} \times \mathbb{O}$, which can be shown to give the gauge theory of the standard model: $U(1) \times SU(2)_L \times SU(3)$. Thus, there is no grand unified theory in terms of gauges that is fundamental (although a GUT may approximately occur at early times cosmologically, at high temperature, where there is conformal flatness). The 'ugly' product gauge that is observed is precisely what is predicted by emanator theory.

(9) Universal thermality is indicated by application of the MIE hypothesis to the choice of whether 'effective' achiral emanation is associative or not. Obviously a restriction to associativity would lose some information transmission capability, but it would then enable the existence of two added layers of complex structure. If we take the MIE in this context to realize the maximum information transmission overall, then the prediction is for effective emanation that is associative (a Clifford subalgebra at this level) but with the two layers of complex structure added. One layer of complex structure is associated with thermal behavior, one layer of complex structure is associated with quantum renormalization behavior (already exhibited).
(10) In the emanation process there is a clear separation between spinorial elements and manifold elements. Manifold elements include geometry and thermality, have no alpha-perturbation effects, and appear to be part of the 'apparatus' from the perspective of the quantum theory.

2.4 Unit-norm emanation

Examples of unit-norm propagation

Unit-norm propagation is a concept that arises in various fields, including optimization, signal processing, and machine learning. Unit-norm propagation is commonly applied in areas such as signal processing in beamforming and direction of arrival estimation [43]; Machine Learning, in algorithms like sparse coding and dictionary learning, where the learned representations are often constrained to unit norm [44]; and in Computer Vision, for tasks like image recognition and feature extraction.

Unit-norm emanation

For a physical description a unit norm object can be used to represent a system, and by repeated transformation to other unit norm objects, it thereby evolves. Mathematical objects that can effect this 'transformation' simply by the rule of multiplication would be objects like division algebras, ideals, and what I'll simply call projections or emanations. In the universal propagator we have a unit norm trigintaduonion (32D) and perform a right multiplication with a chiral (10D) unit norm 'alpha-step' (defined by a max perturbation α into the 29 free dimensions, given by 32 minus one for each chiral choice, and one for the unit normalization overall). Consider multiplication of a given (starting) trigintaduonion from the right with a chiral bi-sedenion as a 'projection' through the (chiral) step indicated. The repeated application and repeated 'chiral steps' thereby arriving at a path describing a chiral propagation. The resulting universal propagation consists of a 32D unit norm trigintaduonion with propagation via right multiplication using a unit-norm, chiral bi-sedenion, with max-α perturbation.

We thereby arrive at a 'Universe Propagator' that takes on the physics parameters desired (notably the fine-structure constant) and imprints them onto the evolution as seen from the 'internal reference frame' where we reference an object in the 4D spacetime with Standard Model gauge field, and where the standard Lagrangian emerges as the necessary 'propagate-able' structure (where Hilbert space must be complex, not real, quaternionic or octonionic, etc. [45]). From maximum information flow with the constructs, and the required emergent complex Hilbert space (thus complex path integral, thus standard quantum operator formalism) we arrive back at the familiar results with justification of their core mathematical representations (e.g., complex Hilbert space), and now with justification of all parameters, all from the emanation hypothesis.

Unit-norm right product propagation is trivial for the division algebras since norm(XY) = norm(X) × norm(Y). From this it is apparent that we have an automorphism group given by the norm itself (since an automorphism if A(XY)=A(X)A(Y)), and in the case of the octonions this automorphism group is G2 [46]. It can be shown that SU(3) is in G2 [46]. Let's now consider the situation with a higher-order Cayley algebra, the Sedenions, 'S'. We obviously don't have norm(S_1S_2) = norm(S_1) × norm(S_2) in general, as this would then allow S to join the ranks of the division algebras, and it is proven that such don't exist above the Octonions [47]. Can we still have a propagation structure? Is it possible to have a 'base' sedenion for which norm(S_{base})=1, and to have a right propagator (product) sedenion also norm(S_{right})=1, such that norm(S_{base} x S_{right}) =1? The answer is yes (see appendix of [1]), when the sedenion has the (chiral) form of an octonion crossed with a real octonion: S_{chiral} = (O,O_{real}) or S_{chiral} = (O_{real},O). Can we continue this to arrive at a propagation structure on the Trigintaduonions? Again the answer is yes, with the chiral form generalizing off the chiral Sedenion as might be expected: T_{chiral} = (S_{chiral}, S_{real}) or (S_{real}, S_{chiral}) [2]. It is proven that this extension process will go no further [2]. What happens is that due to the chiral form we are still able to re-express all T products (or S) as collections of terms involving tri-octonionic products (which have nice properties as described in [1,2]), and this can no longer occur above the (chiral) trigintaduonion level.

2.5 Finding Alpha (α) in Mathematics of Unit-norm Emanation

Starting with the maximal chiral construct with dimensionality $\{10 \subset 32\}$, we can retain unit norm with non-chiral perturbation δ, as long as $\delta < \alpha$, thereby obtaining a fundamental universal constant (suggestively given the label) α.

Consider maximum "unit-norm" propagation (via right multiplications), e.g., a projection (or 'emanation'), where a hypercomplex 'emanator' has maximum propagation dimensionality ten (a doubly-chiral 10dim subspace of the 32dim space of trigintaduonions). The maximum propagation perturbation allowed from the 10dim space into the embedded 32 dim space is given by the maximum perturbation α for the non-10dim part, where this is taken as the definition of α. Computational efforts to determine the maximal perturbation (while remaining "unit norm") recover the known α [3]. Exploration to high precision indicates a possible fractal limit (as noted in [3]), with possible pattern recurrences as in the Mandelbrot Set on complex numbers. A further complication is that the 32 dim hypercomplex trigintaduonion numbers have also become non-associative (but still retain octonionic sub-space 'braid' rules, which are critical in what follows [48]). Even with their complications, however, the hypercomplex formulation provides a unified theory in the sense that states and their evolution are governed by such numbers and their products (similar to the unified theory of ideals expressed in [49]).

2.5.1 Derivation of $\{\alpha\}$

Physics has a lengthy 'love-hate' relationship with hypercomplex numbers. One of the earliest formulations of electromagnetism by Maxwell involved quaternionic mathematics, and even at that time this relationship was off to a difficult start. As stated by Maxwell in a manuscript on the application to electromagnetism in November of 1870 [50]: "... The invention of the Calculus of Quaternions by Hamilton is a step towards the knowledge of quantities related to space which can only be compared for its importance with the invention of triple coordinates by Descartes. The limited use which has up to the present time been made of Quaternions *must be attributed partly to the repugnance of most mature minds to new methods involving the expenditure of thought ...*" (with emphasis mine). The enthusiasm of Maxwell for use of Quaternionic mathematics did not win over other great physicists of his day, such as Josiah Willard Gibbs and Oliver Heaviside, who discarded the quaternionic mathematics in favor of a new mathematics (vector calculus) that they invented so as to avoid the 'foreign' hypercomplex mathematics. In a biography of Hamilton [51], in a quotation attributed to Gibbs: “My first acquaintance with quaternions was in reading Maxwell's E.&M. where Quaternion notations are considerably used. ... I saw, that although the methods were called quaternionic the idea of the quaternion was quite foreign to the subject.”

The stigma associated with hypercomplex mathematics, and the higher-dimensional physics unification attempts of Maxwell and later Einstein, was still significant decades later when Feynman obtained an unusual proof of the homogeneous Maxwell equations [52-55] in a higher (than 3) dimensional space. Feynman was trying to see if any new theoretical theory would be indicated and the fact that he had obtained a novel new way to explain the existing Maxwell’s equations in higher dimensions was not interesting at the time. The inextricable problems of quantum gravity and the discovery of higher-dimensional string theory, among other things, have changed the focus since that time. It has been shown in numerous papers that the (1,9) dimensional superstring has a natural parameterization in terms of octonions [56-58].

In [59,60] the Dirac and Maxwell equations (in vacuum) are derived using octonionic algebras. In [61] a quaternionic equation is described for electromagnetic fields in inhomogeneous media. In [62], the D4-D5-E6 model that includes the Standard Model plus Gravity is constructed using octonionic fermion creators and annihilators. In [63] octonionic constructions are shown to be consistent with the $SU(3)_C$ gauge symmetry of QCD. It would appear that there are a number of implementations involving hypercomplex numbers that are consistent with the Standard Model. But there is still the question of why bother? What is shown in [1] (and Appendix A) is why the bother might be worth it as the new mathematical formulations recovers simultaneously the quantum theory; the critical constant alpha; and many key relations, such as manifest Lorentz Invariance.

2.5.2 Derivation of $\{\alpha, \pi\}$

The chiral trigintaduonion emanation described in Appendix A gives a precise derivation for the mysterious physics constant α (the fine-structure constant) from the mathematical physics formalism providing maximal information propagation, with α being the maximal perturbation amount (a fractal limit), and π being the maximum amount of overall imaginary component contributing to that maximal perturbation. The maximal imaginary component is hypothesized to be at antiphase, thus

'π' phase angle. Component sums becoming angle sums is an aspect of the analyticity hypothesized for the Emanator theory (where exponential map is used), and will be discussed in detail in what follows. Thus, α can be determined by a (fractal) limit process, and separately, by a maximal information propagation argument, where a relation can be shown to exist with the maximum antiphase amount 'π'. The ideal constructs of planar geometry, and related such via complex analysis, give methods for calculation of π to incredibly high precision (trillions of digits), thereby providing an indirect derivation of α to similar precision.

2.5.3 Derivation of $\{\alpha, \pi, C_\infty\}$

The mathematical portion of this section is placed in Appendix A. Understanding what mathematical problem to solve is the main complication. A discussion, thus, follows.

The number system, or algebra, used to describe a physical system is typically the real numbers, sometimes the complex numbers (to describe wavelike phase information), and, rarely, the quaternionic numbers (to describe rotation and EM interactions). In recent theoretical efforts, attention has also been paid to octonionic numbers to describe Quantum Electrodynamics (QED) and Quantum chromodynamics (QCD) interactions [59-65]. The algebras given by real, complex, quaternionic, octonionic, sedenionic, trigintaduonionic, …., are known as the Cayley-Graves algebras, whose dimensions double at each step, one dimension for real, two for complex, four for quaternionic, etc. Maximal *unitary* propagation occurs with the octonion algebra and no higher (thus 'maximal' propagation, seemingly, only in 8 dimensions). What is actually needed in physics 'propagation' is right multiplication with a unit-norm 'propagator', for example, giving rise to a unit-norm result (then iterating to create a path from the infinitesimal propagator steps). If this is sought instead, then a chiral extension can be made from the octonions into the sedenions, and then again into the trigintaduonions, giving rise to a *maximal 'propagation', or projective emanation, in 10 dimensions* within the 32 dimensional trigintaduonions (as shown in [2-5]).

For Real numbers unit norm propagation is trivial, consisting of multiplying by +1 or -1. For Complex numbers unit norm propagation involves multiplication by complex numbers on the classic unit circle in the complex plane, which reduces to simple phase addition according to rotations about the center of that circle (motions on S^1). For quaternion numbers unit norm propagation is still straightforward since it's still, in the end, a normed division algebra, where N(xy)=N(x)N(y). For the quaternions, instead of motion on S^1, we now have motion on S^3, the unit hypersphere in four dimensions. This still holds true for Octonions, with unit norm still directly maintained when multiplying unit norm objects in general. Now the motion is that of a point on a seven dimensional hypersphere S^7. Sedenions are not normed division algebras, lacking linear alternativity and the moufang loop identities [66], thus multiplication of unit norm objects for sedenions (points on S^{15}) will not, generally, remain unit norm, i.e., will leave the S^{15} space.

The question then arises is there is a sub-algebra or projection in the sedenions, that is not just trivially the octonions, that can still allow unit norm propagation? If this works for Sedenions, what about Bi-sedenions (trigintaduonions) and higher dimensional Cayley algebras? In Appendix A (also [1,2]) it is shown that there are two Sedenion subspaces where the unit norm property is retained. This is found again at the level of the Bi-Sedenions by a similar construction. The results were initially explored computationally [3], then later established in theoretical proofs [2-5]. In those proofs a key step fails when attempting to go to higher orders beyond the bi-sedenions and its sub-algebra propagation.

In the RCHO(ST) hypothesis Physics unification was thought to directly entail propagation in terms of hypercomplex numbers [67] (from Reals thru Trigintaduonions in Cayley sequence). This hypothesis was motivated by Maxwell, Feynman and Cayley, in hopes of being able to directly encode the standard model and statistical mechanics. In the end, this idea was not ambitious enough, as will be clear in what follows. Part of the problem is that to get the 10D propagation formalism entails 'projections', not the more familiar mathematical objects directly giving rise to standard propagation (in a complex Hilbert space). Instead, the standard propagation is part of the emergent (with complex Hilbert space) description, as will described later.

The Feynman-Cayley Path Integral proposed in [2] involved use of chiral trigintaduonions in an effort to identify a mathematical framework within which to have a unified propagator theory (and maximal information propagation was sought for such a hypothesized propagator). At its root, this is a hypothesis for an algebraic reality, with algebraic elements describing 'reality' and algebraic multiplicative processes underlying propagation. All of the different 'paths' of propagation are then brought together in a sum – where stationary phase is selected out and the variational calculus basis for much of physics then takes over to offer all of the familiar elegant solutions of classical physics. This is still thought to be the process, but two stages of emergence are indicated: (1) emergence of the emanation (projective) process followed by the (2) emergence of standard propagation in a complex Hilbert space. So, even though we start with RCHO(ST) with the emergence of *emanation*, we end with a framework for emergence of standard propagation with complex propagator in a complex Hilbert space.

With the Feynman-Cayley construction there is a sum on all algebras, with selection for the highest order unit norm propagating algebra. It is shown that the highest order propagating structure is the ten dimensional (10D) unit-norm trigintaduonions elements, that are used here, that are (chiral) extended sedenions that are themselves made from chiral extended octonions.

In quantum physics unitary propagation is a standard part of the description. Efforts to move to algebras to describe such propagation leads to formulations based on the normed division algebras (real, complex, quaternion, and octonion). In an effort to achieve maximal information propagation we relax the unitarity condition and show that multiplication (right) on a unit norm trigintaduonion base by a unit norm chiral trigintaduonions emanator results in a new unit norm product [2]. A path is comprised of repeated (right) multiplications. Each step of the 'emanation' arrived at is a multiplication by a 10D chiral trigintaduonion. Use of methods from noise budget analysis, a constructive perturbation analysis, as well as analysis relating to maximal perturbation according to the Kato Rellich theorem, show that the chiral trigintaduonion with maximal perturbation (outside the 10D into surrounding 32D) has magnitude α, precisely the fine structure constant. A relation between α and π results. Suppose repeated achiral emanation steps can be described as an iterative mapping, with unit-norm constraint resulting in a quadratic relation on components, we then expect the Feigenbaum universal bifurcation parameter, C_∞, to appear according to the number of independent dimensions in a chiral trigintaduonion emanation step and the precise form of the "emanator" construction. The number of effective dimensions is shown to be 29 plus a little more, and a relation between α, π and C_∞ results that is in agreement with the choice of emanator examined in computational studies shown here. The computational studies with the emanator are also explored via "random walks" in the trigintaduonion space during emanation and to explore noise additivity effects. We will show component-level evolution that behaves like a random walk, with random walk asymptotics (established computationally). This will help to establish that the Emanation process is Martingale, since random walk processes are Martingale.

Just from the propagation structure on one path we already see core emergent structure that results in a universal emanation with structural parameters 10,22,78,137 and perturbation maximum α=~1/137. The central notion in the universal emanation hypothesis is that there should be *maximal information flow*, where this is accomplished by finding the highest theoretical dimensionality of unit-norm 'propagation', here called an emanation, which turns out to be 10, then add the maximal perturbation that still allows unit-norm propagation, where that perturbation is into the space the 10D motion is embedded in, here a 32 dimensional (trigintaduonion algebra) space.

2.5.3.1 Maximum information at the edge of chaos

"Maximum information at the edge of chaos" is a concept that arises from the intersection of complexity theory, information theory, and systems theory. It suggests that systems operating at the "edge of chaos"—a state between order and disorder—exhibit the highest levels of complexity and adaptability. The intuition on "edge of chaos" is guided by examination of period doubling phenomena that are observed in some systems as they approach a chaos regime (when adjusting some system parameter, or simply choice of initial conditions). Such chaotic system are universal, with universal properties [68-70].

"Edge of Chaos" is a term that refers to a critical state in dynamic systems where they are neither completely ordered nor completely chaotic (the limit of the period doubling system for example) [71]. Systems at this edge can adapt and evolve, making them more resilient and capable of responding to changes in their environment. It is hypothesized that "at the edge of chaos" systems can maximize their information processing capabilities [72]. This is because they can explore a wide range of possibilities while still maintaining some structure, allowing for the emergence (or capture) of new patterns and behaviors. Many natural and social systems, such as ecosystems, economies, and neural networks, are considered complex adaptive systems. These systems thrive at the edge of chaos, where they can optimally learn from their environment and adapt to new challenges [73]. Further exploration of the concepts of order and complexity are in [74] and [75].

2.5.3.2 Fractal Echo

In Kabbalistic thought, the universe is often seen as a reflection of divine structures, where the microcosm reflects the macrocosm [76,77]. This idea resonates with the concept of fractals, where each part of a fractal pattern mirrors the whole. Kabbalah teaches that the universe is structured in a way that reflects divine attributes (Sefirot). Each Sefirah can be seen as a fractal, where the qualities of the divine are mirrored in the physical world. The concept of Ein Sof (the Infinite) in Kabbalah suggests that the divine essence permeates all levels of existence. This can be likened to fractals, where the infinite complexity arises from simple rules. Kabbalistic texts often use symbolic language and patterns to convey deeper truths. The repetition of themes and symbols is seen as a fractal echo of spiritual truths.

2.6 The Fundamental, Chiral, Emanation Process (non-thermal, regularized)

A 'deeper' phase of universal evolution is described by a theory of emanations, where mathematically invariant emergent structures appear:

$$\boldsymbol{emanation \rightarrow propagation \rightarrow trajectory}$$

At the emanation to propagation emergence, one of the emergent constructs is the familiar path integral based on standard (unitary) propagators in a complex Hilbert space.

We have α, 10,22,78,137 as parameters resulting from analysis on a single path maximal information flow construct, where the number 22 corresponds to the number of emergent parameters in the description of the propagating construct (exact derivation of the 22-parameter to follow). In addition, the time choice is emergent via a multi-path construct, along with the propagator construct, and is coupled in both time step and imaginary time increment. The formulation is inherently embedded in a higher dimensional complex space, thus all of the QFT complex analysis analyticity methods are valid as the assumptions made are now part of the maximal information flow emergent construct.

2.7 The Full (achiral) Emanation Process

The math and math discussion are placed in Appendix A.

2.7.1 Universal Thermality

There is a fundamental complex structure that still remains from the Emanator formalism upon projection from the higher-dimensional Cayley algebras into the maximum propagation described, for the alpha maximum-perturbation trigintaduonion. That complex structure is ‘contact’ analytic and is realized under conditions where limits involving that structure are taken to zero (or to some fixed value at component-level). Two main examples where this has been most significant are in (i) shifting a QFT to a thermal QFT by shifting to imaginary time related to the inverse temperature [66]; and (ii) in the use of dimensional regularization to renormalize QED and QCD [67]. Thus, the process of emanator theory settling on the maximum information propagation dimension is one where higher order complex structure (from non-propagate-able dimensions) is still accessible for regularization processes. A possible *origin* for the complex structures is described in a later section.

2.7.2 Tarot completeness

The Tarot deck has been examined as a narrative generator in various ways, particularly in the fields of psychology, literature, and game design. The concept of "completeness" in this context refers to the idea that the Tarot can generate a ‘complete’ range of narratives or stories based on its archetypal imagery and symbolism. The Tarot is often viewed through the lens of archetypes, which are universal symbols and themes that resonate across cultures [78]. In Jung's work on archetypes and the collective unconscious we can see how Tarot cards can evoke psychological narratives, where each card in the Tarot deck represents specific archetypes that can be woven into narratives.

The structure of the Tarot deck, with its Major and Minor Arcana, can be seen as a storytelling framework. The progression through a selection of the cards can represent a hero's journey or other narrative structure. Consider [79]. Campbell's exploration of the monomyth or hero's journey can be applied to the narrative potential of the Tarot, where each card might represent a stage in the journey (one of many interpretations).

In game design and interactive storytelling, Tarot cards have been used as tools for generating narratives. Players draw cards to create unique storylines, emphasizing the randomness and variability of the narrative [80]. Tarot has also been employed in therapeutic settings to facilitate storytelling for purposes of self-exploration and divination [81].

2.8 The Standard Model, Standard Cosmology, Implications

It is hypothesized that standard model physics with path integral propagation, choice of time, and its assortment of fundamental constants, is emergent from maximal information emanation via trigintaduonions and this will be explored in what follows. Thus, in Emanator Theory, the form of quantum propagation is itself emergent, and within that construct, there is then emergent the functional optimization that describes how the system behaves, e.g., the Lagrangian and choice of time is part of that latter emergent step. Thus, Lagrangians originally introduced as a convenient mathematical constructs, in later physics become endowed with their own physicality, especially in conjunction with the path-integral description to properly capture topological features (the Aharanov-Bohm experiments [82]), are here seen as direct mathematical encapsulations of the fundamental emergent nature of the physical system.

The chiral trigintaduonions T, with right product operation ((T x T) x T)…, used for maximum information transmission, is shown to be H×O when arranged for achiral emanation. When considering a sum over chiral emanations to obtain an achiral emanator, with T as phase factor, we have the exponentiation operation exp(iT), which leads to a theory that is C × H × O. As such, we have the foundation for the associative operator algebra of the Standard Model: $U(1) \times SU(2)_L \times SU(3)$ [48]. A complication with T products is you can have zero divisors. A framework is adopted to remove the zero divisors by requirement of maximum domain of analyticity on the log trigintaduonion multiplication, resulting in a description for the meromorphic precipitation of matter [1,8]. In this process a fundamental quantum is indicated from the zero-divisor residue terms. Analyticity in the form of a Wick rotation also provides a mechanism whereby we can transition to a dimensionful action and quantum and arrive at an explanation for the critical 'smallness' of Planck's constant.

Getting an associative algebra from the repeated operation of a non-associative algebra is first described in [46] in the context of repeated octonion products: ((OxO)xO)…, where the algebra SU(3) can result. This is found to be equivalent to fixing one of the octonion imaginary components in such a repeated-product operation [46,48]. Dixon shows in [48] that the C × H x O product algebra lays the foundation for the associative operator algebra of the Standard Model: $U(1) \times SU(2)_L \times SU(3)$. In later work this is explored in the form of ideals [49].

2.9 The 'Alpha numbers'

The 'Alpha numbers' from mathematical physics theory and from physics observation agree. The agreement is not only to the highest precision in the experimental determination of g-2, for example, but there is also agreement in enumeration and sub-enumeration (groupings) of the constants of the theory. This agreement of the physics observation of the 'alpha mystery', and the aforementioned mathematical physical description of maximal information emanation, suggest an emergent construct (being maximal) and it might thus occur, naturally, in a variety of other (non mathematical physics) scenarios, such as will now be explored.

Chapter 3. Multitudinous Fractals Incarnate

In this chapter fractals, complexity, long-term cycles , and long-range correlations will be discussed. In the current applications to AI/Large Language Models (LLMs), the ability to capture (extreme) long-range correlations (with massive data and massive model) leads to LLMs achieving a new threshold of capability. The LLMs exabit a reasoning 'aha moment' [83] in their generative outputs that appears, to an average human observer, as a reasoning step that an adult human might make. Once a sequence of reasoning steps can be implemented (as in [14, 15, 84]), then we have 'thinking'. This, in turn, forms the basis for similar 'breakthrough' capabilities in any sequential communications medium, including video and audio, but also including anything an autonomous AI Agent might do, in any domain, including robotics.

Fractals, and fractal-like behavior (self-similarity, scaling laws), will occur in a variety of settings. Long-term cycles will be found to either be fractal in some sense, or express system parameters in terms of the Alpha numbers. Shakespeare appears to have had a vision of fractal infinity in the following quote (*Macbeth*, Act 2, Scene 2), from which this chapter title is obtained:

> "I am in blood
> Stepped in so far that should I wade no more,
> Returning were as tedious as go o'er.
> Strange images of death,
> I see in my mind's eye.
> Multitudinous red seas incarnadine,
> ...."

3.1 Alpha from stable, high complexity, long-term cycles

High-complexity long-term cycles refer to intricate and interrelated processes that occur over extended periods, often involving multiple systems and feedback mechanisms.

The Sun is very Stable

The Sun is considered to be a relatively stable star compared to many others in our galaxy, particularly when compared to more massive stars and certain types of variable stars [85]. The Sun is classified as a G-type main-sequence star (G dwarf), specifically G2V. This classification indicates that it has a moderate mass and temperature, which contributes to its stability. In contrast, more massive stars (like O and B types) have shorter lifespans and can undergo dramatic changes, such as supernova explosions [86].

The Sun is approximately 4.6 billion years old and is expected to remain in its highly stable main sequence phase for about another 5 billion years [87]. This long lifespan is indicative of stability, as it allows for a relatively constant output of energy and light over geological timescales. In contrast, more massive stars may evolve rapidly and undergo significant changes in their structure and energy output. While the Sun does exhibit solar activity, such as sunspots, solar flares, and coronal mass ejections, these phenomena are comparatively minor and occur on relatively predictable cycles (approximately 11 years for the solar cycle if ignoring the solar magnetosphere pole flipping).

Comparison with Other Stars [88]

- **Red Dwarfs:** These stars (M-type) are the most common in the galaxy and are known for their long lifespans and low activity levels. However, they can exhibit flares that may be more intense than those of the Sun.
- **Massive Stars:** Stars like Betelgeuse (a red supergiant) are much more massive than the Sun and have much shorter lifespans. They undergo rapid changes and can become unstable, leading to supernova events.
- **Variable Stars:** Certain types of variable stars, such as Cepheid variables, undergo significant changes in brightness due to pulsations in their outer layers, indicating instability.

Galactic Context

In the context of the Milky Way galaxy, the Sun is located in a relatively quiet region of the Orion Arm, away from the more chaotic environments found in the galactic center or in regions with high star formation rates. This location further contributes to its stability and the relatively stable conditions of the solar system.

3.1.1 The Earth-Sun magnetosphere cycle (22 Earth cycles) [89]

The Earth-Sun magnetosphere cycle, also known as the solar magnetic cycle, lasts approximately 22 years. It consists of two 11-year solar cycles, during which solar activity, including sunspots, solar flares, and coronal mass ejections, varies. The magnetic polarity of the Sun reverses at the end of each cycle, influencing the Earth's magnetosphere and space weather. This cycle affects satellite operations, communication systems, and climate patterns on Earth. This cycle is particularly interesting since it exhibits 22 different annual Earth-Sun joint magnetosphere states, in a repeating cycle ongoing for billions of years. The Alpha number 22 is notable in this context, but not all long-term cycles, critical to climate for example, exhibit such parameters in any obvious way, but listed here just the same:

Milankovitch Cycles [90]
Milankovitch cycles are long-term variations in Earth's orbit and axial tilt that affect climate over tens of thousands to hundreds of thousands of years. These cycles include changes in eccentricity, axial tilt, and precession, which influence the distribution of solar energy received by Earth, leading to glacial and interglacial periods.

Glacial-Interglacial Cycles [91]
These cycles refer to the alternating periods of glaciation and interglacial warmth that have occurred over the past 2.5 million years. They are driven by Milankovitch cycles and are characterized by significant changes in global temperatures, sea levels, and ecosystems.

Atlantic Meridional Overturning Circulation (AMOC) [92]
The AMOC is a large system of ocean currents in the Atlantic Ocean that plays a crucial role in regulating climate. It operates on timescales of decades to millennia and is influenced by factors such as salinity, temperature, and ice melt. Changes in the AMOC can lead to significant climate shifts, including periods of warming or cooling in the North Atlantic region.

3.1.2 Observations of 22

The Babylonian 22 year calendar cycle

The Babylonian calendar is an ancient lunar calendar used by the Babylonians, which was based on the cycles of the Moon. It is notable for its complexity and its influence on later calendar systems. The Babylonian calendar consisted of 12 lunar months, each lasting either 29 or 30 days, depending on the sighting of the new moon. This resulted in a year of approximately 354 days. To align the lunar calendar with the solar year (which is about 365.25 days), the Babylonians added an extra month (intercalary month) periodically. This practice was essential for agricultural and religious purposes, ensuring that festivals and seasonal activities occurred at the appropriate times.

Metonic Cycle

The Babylonians developed a system that closely resembles the Metonic cycle, a 19-year cycle used in other ancient cultures, but they extended it to a 22-year cycle. The metonic cycle included 7 intercalary months, resulting in a total of 235 lunar months over 19 years, which is approximately equal to 19 solar years.

The 22-year cycle allowed the Babylonians to synchronize their lunar calendar with the solar year more effectively. By adding intercalary months, they ensured that the lunar months would align with the seasons over the long term.

3.1.3 Fractal G, Scale Invariance, Renormalization Group [1]

The following discussion draws from Chapter 9 in Book 7 in my Physics Series [1], further details can be found there.

The Dirac Large Number Hypothesis [93] has been a mystery that has been noted by Dirac and many other physicists [94-104]. At issue is the odd occurrence of a lot of ratios of 'fundamental' lengths in the theory, that tend to group in families where their ratio is 10^40, or 10^60, or 10^80, etc. (depending on how you want to group the terms). This 'fractal' behavior at large scale (a classic fractal trait being repeating structures at different scales) has not been couched as a fractal phenomenon previously because there wasn't a context to warrant such a supposition. But here we know the maximal information emanation hypothesis is likely to force the emanation process to operate at the edge of chaos, where the process becomes fractal or has a well-defined fractal limit (given the Martingale property [10]), such that we arrive at the fractal G hypothesis, which can be used to explain the origin of the gravitational constant.

From Appendix A we have the hypothesized relation [1,7]:

$$\alpha^{-1} = \left(\sqrt{C_\infty}\right)^{29^*}, \quad where\ dim. = 29^*.$$

Here we are imagining maximum information flowing via 29* effective dimensions, where the phenomenology is concerned with the *number* of dimensions and arrives at a dimensionless parameter α. We might as readily ask for the maximum information that would flow through the 137* independent tri-octonionic terms at chiral-trigintaduonion component level (see appendix A). Here the phenomenology is concerned with the amount of 'noise' that each terms could carry to arrive at a dimensionful parameter 'X':

$$X^{-1} = \left(\sqrt{C_\infty}\right)^{137^*}$$

Evaluation leads to

$$X = 8.05077 \times 10^{-11}[dimensionful\ unit].$$

In SI units the gravitational constant G is:

$$G = 6.674 \times 10^{-11}[N\ m^2\ Kg^{-2}],$$

so, is the nearness of these numbers a coincidence? Taken alone, yes certainly, but taken together with a collection of other odd 'coincidences', noted over the past 100 years [94-104], maybe there's something to this.

Consider relation that motivated Dirac to suggest the large number hypothesis. Working with 'constants' that are seen in the micro and macro scale, such as the Hubble constant H, with age of the universe $\sim H^{-1}$, the mass of the nucleon (proton) m_p, we have the following:

$H^{-1} \cong 4.3 \times 10^{17}$: *the age of the universe* (macro-scale constant)

$\tau = \frac{e^2}{(m_e c^3)} \cong 10^{-23}$: *the strong time scale* (micro-scale constant)

where m_e is the electron mass, c is the speed of light, e^2 is the electrostatic coupling between a single electron and proton, for example, e.g., exhibiting electrostatic force e^2/r^2. The gravitational force between an electron an proton, on the other hand, is $Gm_e m_p/r^2$:

$\frac{Gm_e m_p}{r^2}$: *gravitational force* (macro-scale)

$\frac{e^2}{r^2}$: *electrostatic force* (micro scale)

Consider another ratio of 'fundamental' constants operational on some system that spans the micro and macro realms (e.g., the Universe vs internal, or the macro gravitational system versus its micro system of hydrogen). Consider the mass of the observable Universe M in a ratio with the nucleon mass, which gives $\sim 10^{80}$, thus the square root is used to group it with the $\sim 10^{40}$ ratios. There are going to be obviously going to be large number ratios if we take "universe" / "elemental particle" in some operation, so this isn't what is surprising. What is odd is that these large numbers ratios appear to group as $\sim 10^{40}$, $\sim 10^{60}$, and $\sim 10^{80}$. For the above cases we have:

$$\sim 10^{40} \approx \frac{H^{-1}}{\tau} \approx \frac{e^2}{Gm_e m_p} \approx \sqrt{\frac{M}{m_p}} \, .$$

If we take the Dirac relations as part of the fractal nature of reality (with optimization pushing to "the edge of chaos" and inducing fractal effects), with repeating structure at different scales and similarity relations between 'things' at different scales a possibility. Given this perspective, interpreting the constant ***'G' as a multiscale fit parameter*** across all of these domains, such that the single observed G suffices, is an interesting prospect. Note that not all the terms in the ratios involve G so we can solve for G, approximately, accordingly. Thus, the parameter we know as G is hypothesized to actually be a fractal fit parameter.

3.2 System properties from long-range correlations

The following discussion draws from Chapter 5 in Book 6 in my Physics Series [34], further details can be found there.

Let's now consider the Statistical Mechanics of a system that changes phase as a function of temperature, pressure, or other intensive variable. We will first examine the one-dimensional Ising model with no external field – even though it does not exhibit a phase transition it is foundational to more complex Ising models to follow that do have phase transitions.

The examination of the two-dimensional Ising model starts by reusing the results from the one-dimensional analysis. The Onsager solution for the two-dimensional Ising model [105] will be briefly described using modern notation and techniques [106,107]. The key to Onsager's solution, or statistical cluster analysis in general, or to renormalization group methods to follow in Section 5.12, is a system equation, such as a partition function or Hamiltonian, that has a recalling self-similarity relation (which defines the renormalization group). The Onsager solution will be given exactly, and then approximately solved using renormalization group methods that will give a reasonably close answer to the Onsager solution with significantly less effort. Qualitatively the system examined will have the same phase-transition behavior at a critical temperature. Quantitatively, the system will have the same critical exponent behavior upon approach in a system parameter to the critical value (for the phase transition), e.g., there is universal behavior in critical exponents revealed regardless of approximation as long as that approximation sufficiently captures long-range coupling effects. These subtleties carried over to the context of statistical learning with LLMs indicate that the breakthrough in LLM 'reasoning' might be possible with sufficient scaling, and this is precisely what is observed [14,15].

3.2.1 The Ising Model (this section is at graduate level in mathematical physics)

The section that follows is described in much more detail in [106,107]. Even so, it is an advanced physics discussion so to get the gist simply ignore the math equations, or simply skip to Sec. 3.2.2.

Recall that the Hamiltonian for two particles interacting according to their mutual alignment can be expressed in terms of their inner product:

$$H = -c \sum_{N.N.} s_\alpha \cdot s_\beta \,,$$

Where "$N.N.$" refers to the Nearest Neighbor spin interactions in the sum only. Recall that the local spin operators s_α do not commute with H and that this is often resolved by performing measurements (quantum state preparation) of spin in the z-direction. The focus is then on representation where the $(s_\alpha)_z \rightarrow \sigma_i = \pm 1$ representation is diagonal. Similarly, here, we will add a magnetic field along the z-direction, which results in addition of a Zeeman term to the energy:

$$H = -J \sum_{N.N.} \sigma_i \, \sigma_j - h \sum_{i} \sigma_i \,,$$

where h is proportional to the magnetic field (this can always be set to zero to get the case of no magnetic field).

Let's now explore the temperature behavior of the system where a simple mean field approximation is used (implied) in the form of assuming an overall system magnetization, m, exists (as observed in many natural settings) and describe that system magnetization as the expectation on σ_i:

$$m = \langle \sigma_i \rangle \,.$$

Thus an "order parameter", magnetization, is hypothesized to exist – this is a distinctive characteristic of mean field theory methods.

Writing the mean field system Hamiltonian as a function of a particular spin, here denoted σ_0, we have:

$$H(\sigma_0) = -\,\sigma_0 \left[J \sum_{N.N.} \sigma_j + h \right] = -\,\sigma_0 [JNm + h], \quad N = \sum_{N.N.} 1$$

Using H we can now directly determine $m = \langle \sigma_i \rangle = \langle \sigma_0 \rangle$, an establish a constitutive relation for determining m:

$$m = \langle \sigma_0 \rangle = \frac{Tr[\,\sigma_0 \exp(-\beta H(\,\sigma_0))]}{Tr[\exp(-\beta H(\,\sigma_0))]},$$

which is solved to give:

$$m = \tanh[\beta(JNm + h)]\,.$$

Which is then solved numerically to get a general solution.

For the one-dimensional Ising model the hypothesis of (spontaneous) magnetization given by writing $m = \langle \sigma_0 \rangle \neq 0$ means that there is a critical temperature above which the magnetization is zero and below which the magnetization grows, with two solutions (upward or downward). Near this critical temperature (from below) the above expression for magnetization can make use of Taylor expansion on the tanh, and the small magnetization would then be approximately:

$$m \propto \left(\frac{T_c}{T} - 1\right)^{\frac{1}{2}}.$$

There is, thus, power law behavior (consistent with Landau theory) and the power law, or 'critical', exponent is ½ in this approximation. Thus, if $0 < T_c < \infty$, then we will have a phase transition, and

$$T_c = JN/k_B$$

as indicated by the Bragg-Williams approximation where the spins are taken to be independent. This turns out to be incorrect, however, as $T_c = 0$, as will be shown in exact derivations, and again later using renormalization group methods. The discrepancy arises because the mean field and Bragg-Williams approximations have effectively arrived at a description with infinite-range interactions, and this is what is permitting the false phase transition to appear. Furthermore, the mean-field derivation would suggest that the number of nearest neighbors is all that matters and not lattice structure or dimensionality, both very wrong in practice. So, knowing the limits of mean field applications will be necessary.

In [108,109], a very clever approach to testing the low temperature regime of the one-dimensional Ising model can be made. Let's suppose that the system is in such a low energy state that there is only one transition between blocks of spin. Thus, for $i \leq l$ we have spin up, and for $i > l$ we have spin down. For length N Ising chain (with free ends) there are $N - 1$ such states. At temperature T the free energy change due to these fluctuations is:

$$\Delta G = 2J - k_B T ln(N-1)$$

which is always negative for large N, thus, the magnetization should be zero. A similar argument can be done for the two-dimensional Ising problem, where now the lowest order fluctuation is a separating line between two regions of block spin (one up and one down). The boundary length L between the two spin regions (counted according to the number of spins in relevant, anti-aligned, nearest neighbor sums) then allow the change in energy in the domain wall (over zero ground state) to be simply written as:

$$\Delta E = 2LJ.$$

The number of possible boundary walls of length L goes as 2^L (ignoring edge effects). For a $N \times N$ lattice there are N stating points for the boundary on the left, and there are N steps across, but half the time the boundary moves transversely, thus the average length L is $2N$. Putting this together we have for free energy change due to boundary wall fluctuations:

$$\Delta G = 4NJ - k_B T ln(N 2^{2N}).$$

This system is stable against forming the domain indicated if less than a critical temperature, where:

$$T_c = \frac{2J}{k_B ln2} = 2.885 \frac{J}{k_B}.$$

This is remarkably close to the exact answer due to Onsager:

$$T_c = 2.269 \frac{J}{k_B}.$$

1-D Ising Model – Exact Solution

Let's now obtain the exact solution to the 1-D Ising Model:

$$H = -J \sum_{i=1.} \sigma_i \, \sigma_{i+1} \, .$$

We obtain the partition function:

$$Z_N = \sum_{\sigma_1=\pm 1} \dots \sum_{\sigma_N=\pm 1} e^{\beta J \sum_{i=1} \sigma_i \, \sigma_{i+1}}.$$

Making use of

$$\sum_{\sigma_N=\pm 1} e^{\beta J \, \sigma_N \, \sigma_{N+1}} = 2 \cosh \beta J$$

we have:

$$Z_N = 2[2 \cosh \beta J]^{N-1}.$$

Now that we have the partition function, we can obtain the free energy as:

$$F = -N k_B T \ln(2 \cosh \beta J) \, .$$

Let's now add a magnetic field as previously but also have our 1-D Ising Chain form a ring to eliminate edge effects (now more complicated with the magnetic field). So,

$$H = -J \sum_{i=1..N} \sigma_i \, \sigma_j - h \sum_{i=1..N} \sigma_i \, ,$$

with $N + i = i$:

$$H = - \sum_{i=1..N} \left\{ J \, \sigma_i \, \sigma_j + \frac{h}{2} (\sigma_i + \sigma_{i+1}) \right\}.$$

This is typically solved by introducing a transfer matrix. To begin, we have

$$Z_N = \sum_{\sigma_1 = \pm 1} \cdots \sum_{\sigma_N = \pm 1} e^{\beta \sum_{i=1..N} \left\{ J \, \sigma_i \, \sigma_j + \frac{h}{2} (\sigma_i + \sigma_{i+1}) \right\}}.$$

Or,

$$Z_N = \sum_{\sigma_1 = \pm 1} \cdots \sum_{\sigma_N = \pm 1} P_{\sigma_1 \sigma_2} P_{\sigma_2 \sigma_3} \dots P_{\sigma_N \sigma_1} = Tr \, P^N,$$

where

$$P_{1,1} = e^{\beta (J+h)}; \; P_{-1,-1} = e^{\beta (J-h)}; \; P_{-1,1} = P_{1,-1} = e^{-\beta J}$$

If we diagonalize P in terms of eigenvalues $\lambda_{1,2}$ and then evaluate $Tr \, P^N$ we get:

$$Z_N = (\lambda_1)^N + (\lambda_2)^N,$$

where

$$\lambda_{1,2} = e^{\beta J} \cosh \beta h \pm \sqrt{e^{2\beta J} \sinh^2 \beta h + e^{-2\beta J}}.$$

The free energy is now determined to be:

$$F = -k_B T \ln((\lambda_1)^N + (\lambda_2)^N) = -k_B T \left\{ N \ln \lambda_1 + \ln \left[1 + \left(\frac{\lambda_2}{\lambda_1} \right)^N \right] \right\}.$$

Since $\lambda_1 > \lambda_2$ we have in the thermodynamics limit as $N \to \infty$ that:

$$F = -N k_B T \ln \left(e^{\beta J} \cosh \beta h + \sqrt{e^{2\beta J} \sinh^2 \beta h + e^{-2\beta J}} \right)$$

We can now evaluate the magnetization:

$$m = \langle \sigma_0 \rangle = - \frac{1}{N} \frac{\partial G}{\partial h} = \frac{(\sinh \beta h)}{\sqrt{\sinh^2 \beta h + e^{-4\beta J}}}.$$

This is the exact solution and it shows that as $h \to 0$ there is no spontaneous magnetization (thus no phase transition).

1-D Ising Specific Heat

Using the exact solution to the one dimensional Ising problem with

$$F = -k_B T \ln Z = -NK_B T \ln \lambda_m$$

where λ_m is the larger of the two eigenvalues

$$\lambda_\pm = e^K \cosh B \pm \{e^{2K} \sinh^2 B + e^{-2K}\}.$$

Let's derive an expression for the specific heat in zero field, $C_{H=0}$, in the limit where $T \to 0$. Also, show that the field magnetic susceptibility $X_{H\to 0}$ diverges in the limit where $T \to 0$. What is the nature of this singularity?

Partial Answer

We have

$$F = -K_B T \ln Z = -Nk_B T \ln\left\{e^4 \cosh B + (e^{2k} \sinh^2 B + e^{-2k})^{1/2}\right\}$$

Thus

$$\frac{J}{k_B T} = k \;; \frac{\beta H}{k_B T} = B.$$

Using the relations

$$C_{H=0} = T\left(\frac{\partial S}{\partial T}\right), \qquad \left(\frac{\partial F}{\partial T}\right) = -S$$

We start with:

$$F_{H=0} = -Nk_B T\{e^k + e^{-k}\} = -Nk_B T \ln\{2\cosh k\}$$

$$\left(\frac{\partial F}{\partial T}\right) = -Nk_B T \ln\{2\cosh k\} - Nk_B T\left(\frac{2\sinh k}{2\cosh k}\right)\left(-\frac{J}{k_B}\frac{1}{T^2}\right)$$

$$= -Nk_B T \ln\{2\cosh k\} + NJ(\tanh k)\left(\frac{1}{T}\right)$$

Thus

$$S = Nk_B \ln(2\cosh k) - NJ(\tanh k)\left(\frac{1}{T}\right)$$

and

$$\left(\frac{\partial S}{\partial T}\right) = Nk_B(\tanh K)\left(-\frac{J}{k_B T^2}\right) - NJ\tanh k\left(\frac{-1}{T^2}\right) - \frac{NJ}{T}(\mathrm{sech}^2 k).$$

So

$$C_{H=0} = -NJ\,\mathrm{sech}^2\left(\frac{J}{k_B T^2}\right) \to \quad \lim_{T\to 0} C_{H=0} = 0$$

1-D Ising Solution by constructing entropy $S(E,N)$

Let's solve the 1 dimensional Ising problem by direct construction of the entropy function $S(E,N)$. Use this entropy function to derive the Helmholtz free energy $F(T,N)$ and show that result agrees with the result obtained by summing the partition function Z.

$$F = E - TS$$

$$E_m = -2NJ + 2mJ \;\; ; S(m) = k_B \ln\left[\binom{N}{m}\right]$$

$$F = (2J)(m - N) - Tk_B \ln\left[\frac{N!}{m!\,(N-m)!}\right]$$

$$F = 2J(m - N) - k_B T\{\ln N! - \ln m! - \ln(N - m)!\}$$

$$F = 2J(m - N) - k_B T\{N \ln N - m \ln m - (N - m)\ln(N - m)\}$$

The derivation obtained by summing the partition function Z has the Hamiltonian going from $-NJ$ to NJ as we go from order to disorder. Observe that E_m defined thus far is shifted by NJ in that it goes from$-2NJ$ to 0. Lets make the model consistent by making the energy scales consistent, so redefine: $E_m = -NJ + 2mJ$

From Z derivation:

$$F = -Nk_B T \ln\{e^k + e^{-k}\} = -N\left(\frac{J}{K}\right)\ln\{2\cosh k\}$$

$$= -Nk_B T \ln(2\cosh k)$$

For entropy we then get:

$$S(m) = k_B\{N \ln N - m \ln m - (n - m)\ln(N - m)\}$$

$$E = -NJ + 2mJ \rightarrow m = (E + NJ)/2J$$

So,

$$\left(\frac{\partial S}{\partial E}\right) = \frac{1}{T} = \left(\frac{\partial S}{\partial m}\right)\left(\frac{\partial m}{\partial E}\right)$$

$$= \left(\frac{1}{2J}\right) k_B \left\{\frac{\partial}{\partial m}[-m \ln m - (N - m)\ln(N - m)]\right\}$$

$$= \left(\frac{K_B}{2J}\right)\left\{-\frac{m}{m} - (N - m)\frac{(-1)}{(N - m)} - \ln(N - m)(-1) - \ln m\right\}$$

$$= \frac{K_B}{2J}\left(\ln\left(\frac{N - m}{m}\right)\right) = \frac{1}{T}$$

Thus,

$$\frac{N - m}{m} = e^{2k} \rightarrow m = \frac{N}{e^{2k} + 1} \rightarrow \quad m = \frac{2Ne^{-k}}{\cosh k}.$$

Using this we then write:

$$F = \left[-NJ + 2J\left(\frac{2Ne^{-k}}{\cosh k}\right)\right] - Tk_B\left\{N \ln N - \left(\frac{2Ne^{-k}}{\cosh k}\right)\ln\left(\frac{2Ne^{-k}}{\cosh k}\right) - \left(N - \frac{2Ne^{-k}}{\cosh k}\right)\ln\left(N - \frac{2Ne^{-k}}{\cosh k}\right)\right\}$$

$$F = -JN + 2JN - \frac{NJ}{K}\ln(e^{2k} + 1) = JN - \frac{JN}{K}\ln\left[\frac{e^k + e^{-k}}{e^{-k}}\right]$$

$$= -\frac{JN}{k}\ln(2\cosh k) = -Nk_BT\ln(2\cosh k)$$

Evaluation of Magnetic Susceptibility for 1-D Ising Model

Let's evaluate the magnetic susceptibility

$$\chi = \left[\frac{\partial M}{\partial H}\right]_{H\to 0, T\to 0}$$

For the one-dimensional Ising problem and show that χ is singular in the limit where $T \to 0$ (and $H \to 0$)

Solution

By definition:

$$M_I = \langle \sum_{i=}^{N} s_i \rangle = kT \cdot \frac{1}{Z_I} \cdot \frac{\partial Z_I}{\partial H} = -\frac{\partial F_I}{\partial H}$$

where

$$Z_I = \sum_{s_1=\pm 1} \sum_{S_2=\pm 1} \cdots \sum_{S_{N=\pm 1}} e^{\beta V \Sigma s_i \cdot s_j + B \Sigma S_1}$$

where

$$B = \frac{aH}{k_BT}, \qquad \beta = \frac{1}{k_BT}.$$

We have:

$$F_I = -Nk_BTln\lambda_+$$

$$\lambda_+ = e^k\cosh B + \sqrt{e^{2k}\sinh^2 B + e^{-2k}}, \qquad k = \beta V$$

$$M_I = -\frac{\partial F_I}{\partial H} = \frac{Nk_BT}{\lambda_B}\left(\frac{\alpha}{k_BT}\right)\frac{\partial \lambda_+}{\partial B}$$

We need to evaluate $\frac{\partial \lambda_+}{\partial B}$ and $\frac{\partial^2 \lambda_+}{\partial B^2}$:

$$\frac{\partial \lambda_+}{\partial B} = e^k \sinh B + \frac{1}{2}\frac{2\sinh B\cosh B\, e^{2k}}{\sqrt{e^{2k}\sinh^2 B + e^{-2k}}}$$

$$\frac{\partial^2 \lambda_+}{\partial B^2} = e^k\cosh B$$
$$-\frac{1}{4}(e^{2k}\sinh^2 B + e^{-2k})^{-3/2}(2e^{2k}\sinh B\cosh B)^2$$
$$+\frac{1}{2}(e^{2k}\sinh^2 B + e^{-2k})^{-\frac{1}{2}}[2e^{2k}\cosh^2 B + 2e^{2k}\sinh^2 B]$$

So,

$$\chi_T = \frac{\partial M}{\partial H} = \frac{\alpha}{k_B T}\cdot\frac{\partial M}{\partial B} = \left(\frac{\alpha}{k_B T}\right) N k_B T \left[\left(\frac{-1}{\lambda_+^3}\right)\cdot\left(\frac{\partial \lambda_+}{\partial B}\right)^2 + \frac{1}{\lambda_+}\frac{\partial^2 \lambda_+}{\partial B^2}\right]$$

As $H \to 0 \to \frac{\partial \lambda_+}{\partial B} \to 0, \frac{\partial^2 \lambda_+}{\partial B^2} \to e^k + \frac{1}{2}e^k \cdot 2e^{2k} = e^k + e^{3k}$, and $\lambda_+ \to e^k + e^{-k}$. So

$$\chi_T|_{H=0} = \left(\frac{\alpha^2}{k_B T}\right) N \cdot \left[\frac{1}{(e^k + e^{-k})^2}\cdot 0 + \frac{1}{(e^k + e^{-k})}\cdot (e^k + e^{3k})\right]$$

Thus

$$\chi_T|_{H=0} = \frac{N\alpha^2}{k_B T}\cdot e^{2k} = \left(\frac{N\alpha^2}{k_B T}\right) e^{\left(\frac{2V}{k_B T}\right)}$$

And as $T = 0$ we see that $\chi_T \to \infty$.

1-D Ising Model derived from microcanonical ensemble

Let's compute the solution of the one dimensional Ising problem using the microcanonical ensemble. Find an expression for $S(E, H = 0)$. Also find $F(T, H = 0)$.

Solution

$$S(E) = k_B \ln w(E)$$

and $w(E)$ is the density of states. So, for this situation

$$S(E) = k_B ln \frac{2N!}{m!\,(N-m)!} = k_B[ln2 + lnN! - lnm! - ln(N-m)!]$$

Using the Stirling approximation:

$$S(E) \cong k_B[ln2 + NlnN - mlnm - (N-m)ln(N-m)]$$

Now we have the relations $\frac{\partial S}{\partial E} = \frac{1}{T}$ and $E = -(N-2m)V \to \frac{\partial E}{\partial m} = 2V$, so:

$$\frac{\partial S}{\partial m} = \frac{\partial S}{\partial E}\cdot\frac{\partial E}{\partial m} = \frac{2V}{T} = k_B ln\left(\frac{N-m}{m}\right), \quad m = \frac{E+NV}{2V}$$

Thus,

$$E = NV \tanh\frac{V}{k_B T}$$

$$S(E) = k_B\left[ln2 + NlnN - \left(N - \frac{E+NV}{2V}\right) ln\left(N - \frac{E+NV}{2V}\right) - \left(\frac{E+NV}{2V}\right) ln\left(\frac{E+NV}{2V}\right)\right]$$

$$= k_B \begin{bmatrix} ln2 + NlnN - \frac{N}{2}\left(1 + \tanh\frac{V}{k_BT}\right) ln\frac{N}{2}\left(1 + \tanh\frac{V}{k_BT}\right) \\ -\frac{N}{2}\left(1 - \tanh\frac{V}{k_BT}\right) ln\frac{N}{2}\left(1 - \tanh\frac{V}{k_BT}\right) \end{bmatrix}$$

$F = E - TS = \frac{N}{2}\tanh\frac{V}{k_BT} - k_BT[...]$, with above expression for [...].

The 2-D Ising Model – Exact Solution

The exact solution to the 2-D Ising model was obtained by Onsager in 1944 [105]. Even using a modern approach to simplify things as with Schultz [110] and then using a clearly written synopsis by Plischke [106], it is still too much to consider in detail here. For a description of this amazing derivation see Plischke [106]. To have a sense of the maneuvers involved in the mathematics employed by Schultz refer to Chapter 10 where Euclideanized time is described. In that section we see that a common maneuver in quantum field theory in 3+1 Lorentzian spacetime is to shift to imaginary time to get a 4D Euclidean spacetime, where the integrals become well-defined. The Wick rotation to go from real time to imaginary time essentially takes a (N-1)+1 dimensional Lorentzian field theory over to a N dimensional (Euclidean) statistical field theory. This can be done in reverse (Wick rotation the other way) as long as the Lorentzian theory is well-defined. If working in a quantum mechanics formulation (not quantum field theory form), the Lorentzian description will automatically be well-defined. Thus, a maneuver used in quantum field theory (Wick rotation) will be used by Schultz in reverse to take a difficult 2D theory (with angular momentum commutation) over to a much less complicated 1+1D quantum system description (with canonical commutation). The simplified description is then solved to get he critical temperature:

$$\sinh\frac{2J}{k_BT_c} = 1 \rightarrow \frac{k_BT_c}{J} \cong 2.269\,...$$

The 2-D Ising Model has Specific Heat with logarithmic singularity

Using the exact solution to the 2-dimensional Ising problem discussed in Landau [111], let's show that the specific heat exhibits a logarithmic singularity (as a function of temperature) at the critical temperature T_c:

$$\phi = -NT\ln 2 + NT\ln(1 - x^2)$$

$$-\frac{NT}{2(\pi)^2}\int_0^{2\pi}\int_0^{2\pi} \ln[(1 + x^2)^2 - 2x(1 - x^2)(\cos w_1 + \cos w_2)]\, dw_1 dw_2$$

where

$$x = \tanh(J/T)$$

The function $\phi(T)$ has a singularity when the argument of the logarithm can vanish.

The minimum value for a zero in the argument, in terms of $x(T)$, will correspond to the phase transition point. We want the minimum value, k_c, for a zero in the argument of the log. So, we must minimize as a function of w_1 and w_2 of:

$$(1 + x^2)^2 - 2x(1 - x^2)(\cos w_1 + \cos w_2) = 0$$

Which is found to occur when $\cos w_1 = \cos w_2 = 1$, with positive root of interest $x_c = \sqrt{2} - 1$.
What is vital thus far is that $\cos w_1 = \cos w_2 = 1$ provides x_c. Near the transition point not only is the argument of the log no longer considered zero (for the singularity) but we get contributions from the different $w_1' s\ and\ w_2's$. Still, the most important contributions are with those that have $x \sim x_c$, or, in terms of an expansion about the minimum and truncating the rest, we get the approximate integrand:

$$A = (1 + x^2) - 2x(1 - x^2)(2 + w_1^2 + w_2^2)$$

Now if we expand A about x_c, transform $x \to T$ and pick off leading order behaviour we get integrand of the form:

$$A = C_1 t^2 + C_2(w_1^2 + w_2^2)$$

thus

$$\phi \cong \int_0^{2\pi}\int_0^{2\pi} A dw_1 dw_2 = a + \frac{1}{2} b(T - T_c)^2 \ln|T - T_c|$$

and

$$C = \frac{\partial^2 \phi}{\partial T^2} = -bT_c \ln|T - T_c|.$$

Thus, the specific heat exhibits a logarithmic singularity.

1-D Ising Renormalization Group

Recall the partition function for the theory:

$$Z_N = \sum_{\sigma_1=\pm 1} \dots \sum_{\sigma_N=\pm 1} e^{\beta J \sum_{i=1} \sigma_i \sigma_{i+1}}.$$

At the level of the partition function we now perform a very different kind of approximation maneuver. For contrast, recall that in mean field theory we average over all but a few degrees of freedom, an infinity of contributions is reduced to a finite number of degrees of freedom. For renormalization the objective is to remove an infinite number of degrees of freedom, but still leave an infinite number of degrees of freedom. Suppose we remove half an infinite 1D lattice by averaging out every other position. We can then describe a relation between the original partition function and that of the renormalized theory (both being 1D infinite lattices). In doing this we arrived at a renormalization group relation, from which very useful results can be obtained easily.

Let's regroup the above expression:

$$Z_N = \sum_{\sigma_1=\pm1} \dots \sum_{\sigma_N=\pm1} e^{\beta J[\sigma_1\sigma_2+\sigma_2\sigma_3]} e^{\beta J[\sigma_3\sigma_4+\sigma_4\sigma_5]} \dots.$$

and then sum out the even numbered indices:

$$Z_N = \sum_{\sigma_1=\pm1} \dots \sum_{\sigma_{odd}=\pm1} \begin{matrix} \left\{e^{\beta J[\sigma_1+\sigma_3]} + e^{-\beta J[\sigma_1+\sigma_3]}\right\} \\ \times \left\{e^{\beta J[\sigma_3+\sigma_5]} + e^{-\beta J[\sigma_3+\sigma_5]}\right\} \dots \end{matrix}.$$

As indicated above, we want to relate this to the form of he original 1D Ising partition function with half the lattice density. To do so let's rewrite in the form:

$$e^{K[s+s']} + e^{-K[s+s']} = f(K)e^{-K'ss'}, \qquad K = \beta J,$$

for which we have:

$$Z_N = \sum_{\sigma_1=\pm1} \dots \sum_{\sigma_{odd}=\pm1} f(K)e^{K'[\sigma_1\sigma_3]} f(K)e^{K'[\sigma_3\sigma_5]} \dots$$

Thus,

$$Z_N(K) = [f(K)]^{\frac{N}{2}} Z_{N/2}(K').$$

From the latter form, known as a Kadanoff transformation [107,112,113] we can quickly obtain some useful results. To use this form, however, we must verify it exists, i.e., we must obtain the unknowns $f(K)$ and K'.

If $s = s' = \pm1$ we have:

$$e^{2K} + e^{-2K} = f(K)e^{K'}.$$

If $s = -s' = \pm1$ we have:

$$2 = f(K)e^{-K'} \rightarrow f(K) = 2e^{K'}$$

Thus

$$f(K) = 2\cosh^{1/2}(2K)$$
$$K' = 1/2\, ln \cosh(2K)$$

and we see that the transformation exists.

If writing the partition function in relation to free energy we separate a N factor, leaving the remaining factor an intensive variable, independent of system size:

$$\ln Z_N(K) = Ng(K)$$

Thus,

$$g(K') = 2g(K) - \ln\left[2\cosh^{\frac{1}{2}}(2K)\right].$$

The $f, g,$ and K' relations are the renormalization group equations. Successive application of these equations will result in flow to a stable fixed point of the theory (or away from an unstable fixed point). For the 1D Ising result the flow will take the K parameter to zero or infinity, indicating there is no phase transition in the theory.

The 2-D Ising Renormalization Group

For the 2D problem we will repeat the strategy outlined in the 1D case. Starting with the partition function for the theory:

$$Z_N = \sum_{\sigma_1=\pm1} \cdots \sum_{\sigma_N=\pm1} e^{\beta J \sum_{i=1} \sigma_i \sigma_{i+1}}.$$

except now the pairwise couplings occur according to the nearest neighbor in a square lattice configuration. Suppose the four nearest neighbors of σ_5 are σ_1, σ_2, σ_3, and σ_4,we can write:

$$Z_N = \sum_{\sigma_1=\pm1} \cdots \sum_{\sigma_N=\pm1} \cdots e^{\beta J \sigma_5[\sigma_1+\sigma_2+\sigma_3+\sigma_4]} e^{\beta J \sigma_{10}[\sigma_6+\sigma_7+\sigma_8+\sigma_9]} \cdots.$$

where alternating diagonals are separated out (σ_5 and σ_{10} might be on the same diagonal, for example). This is done to then sum out the spins on alternating diagonals in the lattice (upon removal of the alternating diagonals on the lattice you have a new square lattice with half the node density and turned at 45 degrees relative to the original lattice). In this manner we partly sum the partition function to get:

$$Z_N = \sum_{\sigma_1=\pm1} \cdots \sum_{\sigma_{odd\ diagonal}=\pm1} \cdots \times \left\{e^{K[\sigma_1+\sigma_2+\sigma_3+\sigma_4]} + e^{-K[\sigma_1+\sigma_2+\sigma_3+\sigma_4]}\right\} \times \cdots.$$

in a similar manner to before, we now want to obtain the form:

$$e^{K[\sigma_1+\sigma_2+\sigma_3+\sigma_4]} + e^{-K[\sigma_1+\sigma_2+\sigma_3+\sigma_4]} = f(K)e^{-K'[\sigma_1\sigma_2+\sigma_2\sigma_3+\sigma_3\sigma_4+\sigma_4\sigma_1]}.$$

The inequivalent choice of spin configuration now gives rise to four equations, not two, and guessing that a solution exists in the form:

$$e^{K[\sigma_1+\sigma_2+\sigma_3+\sigma_4]} + e^{-K[\sigma_1+\sigma_2+\sigma_3+\sigma_4]}$$

$$= f(K)\exp\left[\begin{array}{c}\left(\frac{1}{2}\right)K_1(\sigma_1\sigma_2 + \sigma_2\sigma_3 + \sigma_3\sigma_4 + \sigma_4\sigma_1) \\ +K_2(\sigma_1\sigma_3 + \sigma_2\sigma_4) \\ +K_3(\sigma_1\sigma_3\sigma_2\sigma_4)\end{array}\right]$$

we find that it does with:

$$K_1 = \frac{1}{4}\ln\cosh(4K), \qquad K_2 = \frac{1}{8}\ln\cosh(4K)$$

$$K_3 = \frac{1}{8}\ln\cosh(4K) - \frac{1}{2}\ln\cosh(2K)$$

$$f(K) = 2[\cosh(4K)]^{1/8}[\cosh(2K)]^{1/2}.$$

Consider the implications of the form now shown to exist by these solutions, it indicates that our nearest neighbor coupling, upon renormalization as indicated, results in a new coupling that is not only nearest neighbor, but also next nearest neighbor, and also has a 4-coupling involving spin couplings that complete a square in the lattice. Notice that if we try to approximate the renormalization equation on the form above by dropping the K_2 and K_3 terms we get:

$$e^{K[\sigma_1+\sigma_2+\sigma_3+\sigma_4]} + e^{-K[\sigma_1+\sigma_2+\sigma_3+\sigma_4]}$$

$$= f(K)\exp\left[\left(\frac{1}{2}\right)K_1(\sigma_1\sigma_2 + \sigma_2\sigma_3 + \sigma_3\sigma_4 + \sigma_4\sigma_1)\right]$$

which is precisely the same form as in the 1D Ising problem, with similar renormalization group flow results, i.e., no phase transition. Thus, to capture the local correlations sufficiently to demonstrate that there is a phase transition we must at least show the theory at complexity greater than the 1D Ising form. So, let's now drop only the K_3 term but let's still approximate the K_2 contribution describing next nearest neighbor interactions as a modification to the nearest neighbor interaction such that we have: $K_1 \to K'(K_1, K_2)$. Thus, we have a very similar renormalization group equation form to the 1D Ising case:

$$Z_N(K) = [f(K)]^{\frac{N}{2}} Z_{N/2}\big(K'(K_1, K_2)\big).$$

If we again let

$$\ln Z_N(K) = Ng(K)$$

we get

$$g(K') = 2g(K) - \ln\left[2[\cosh(4K)]^{1/8}[\cosh(2K)]^{1/2}\right].$$

When all spins are aligned we have the relation

$$K' \cong K_1 + K_2 = \frac{3}{8}\ln\cosh(4K).$$

There is a fixed point at

$$K = \frac{3}{8} \ln \cosh(4K) \rightarrow \quad K \cong 0.50698 \ldots$$

The fixed point is associated with a critical temperature:

$$\frac{k_B T_c}{J} \cong 1.972 \ldots$$

where the exact answer (with a lot more work) is:

$$\frac{k_B T_c}{J} \cong 2.269 \ldots .$$

3.2.2 Examples of scaling in computational models – AI with LLM

From the physics examples above it is clear that to exhibit critical learning of system parameters the data acquisition used in the modeling must retain long-range correlation information. This is not a simple task and was not accomplished efficiently in the case of Large Language Models (LLMs) until the attention mechanism and transformers were brought together in 2017 [14]. Once there was sufficient capture of long-range correlation information (in the text), the language model performance began to exhibit clear power law scaling in performance [15]. In the latter paper, by Kaplan et al, the reduction in the cross-entropy loss function L was examined in how it scaled with increase in compute time C (PetaFlop Days), increase in dataset size D (in tokens), and increase in model parameters P (non-embedding). Here are the scaling relations for decrease in loss L versus increase in C, D, or P:

$$L = \left(\frac{C}{2.3 \times 10^8}\right)^{-0.050}$$

$$L = \left(\frac{D}{5.4 \times 10^{13}}\right)^{-0.095}$$

$$L = \left(\frac{P}{8.8 \times 10^{13}}\right)^{-0.076}$$

Note that in establishing these relations in [15], the power-law relationship for each is established while not "bottlenecked" by the other two. Thus, the optimal performance indicated (smallest Loss) is when all scaling parameters are at their maximum.

Similar scaling behavior is seen in [114]. The key result is that the scaling results provide a predictive framework for future performance (and larger compute investments). Bigger models and bigger data are, thus, likely to offer better performance. Bigger models can be trained using pipelining [115] and are more amenable to parallelization [116]. Sparsity and branching allows for further speedup [117-119]. The model can also grow as it learns for a dynamic optimization [120,121]. Not surprisingly, further increase in compute time with the setup in [15] led to continued scaling with similar power law [122] for an additional two orders of magnitude increase in compute time:

$$L = 2.57(C)^{-0.048}.$$

Thus, long-range correlation in texts [123] allows capture of information and scaling-law performance with increase in compute or model capability in a predictable way, empirically [124]. Further developments, such as pre-training (part of ChatGPT), would allow further scaling behavior in a new domain [125, 126], and again with test-time compute training, so the empirical growth in capability is expected to continue, with improved scaling as other training stages are added, for the foreseeable future.

Chapter 4. The Mystery of Life

Much of the material in this chapter draws from a prior publication: "Data Analytics, Bioinformatics, and Machine Learning", chapters 5,8-10 [127]. Further details, and more applications, are explored there.

4.1 Information encoding molecules, autocycle origins

Autocycles and self-sustaining cycles are important concepts in the study of early life chemistry and abiogenesis, as they provide insights into how simple organic molecules could evolve into more complex systems capable of self-replication and metabolism.

Autocycles are self-sustaining chemical cycles that can maintain themselves through a series of reactions. They are characterized by the ability to regenerate their starting materials, allowing for continuous operation without the need for external inputs. In the context of early life chemistry, autocycles can be seen as precursors to metabolic pathways.

One of the most well-known examples of an autocycle is the Krebs cycle (or citric acid cycle), which is a series of enzymatic reactions that produce energy through the oxidation of acetyl-CoA derived from carbohydrates, fats, and proteins. However, in the context of abiogenesis, simpler autocyclic reactions are needed.

Early Earth conditions likely facilitated the formation of simple organic molecules through processes such as seen in Miller-Urey experiments [128], which simulated primordial conditions. These reactions could lead to the formation of amino acids, nucleotides, and other building blocks of life. The concept of self-sustaining cycles can then be extended to formation of early metabolic networks that could have emerged in prebiotic environments [129]. For instance, the formose reaction can produce sugars from formaldehyde, which could then participate in further reactions to form nucleotides and amino acids.

The debate between "metabolism-first" and "genes-first" hypotheses in abiogenesis often involves discussions of autocycles. The metabolism-first hypothesis suggests that self-sustaining metabolic cycles could have existed before the emergence of genetic material, providing a metabolism first framework for the evolution of life [130]. To put this issue in perspective, it is important to understand what the emergence of genetic material might look like, so that is briefly described next. It is also important to understand the abiogenesis setting, so that is also described.

RNA World Hypothesis – A Preview

The RNA world hypothesis posits that self-replicating RNA molecules were precursors to current life forms. Autocatalytic cycles involving RNA could have led to the development of more complex biochemical systems. For example, certain RNA molecules can catalyze their own replication, creating a self-sustaining cycle of replication and evolution. Still, having sufficient RNA as the basis for autocatalytic systems would be difficult to envisage in an abiogenesis context. The precursor to RNAs in the form of PAHs (Polycyclic Aromatic Hydrocarbons) more generally can be envisaged, however, organized in a naturally occurring circumstance, the undersea thermal vent, where a hydrocarbon processing stack can occur with a long established high chimney separation columns (with PAH supplementation from cometary sources as well). The tall vent 'separation column' would have been taller on Mars, just like the peaks are taller due to the lower gravity.

The RNA World Hypothesis is a widely discussed theory in the field of molecular biology and evolutionary biology. RNA molecules capable of both storing genetic information and catalyzing chemical reactions, which would allow them to play a central role in the origin of life. The key attributes of RNA as a basis for life are:

1. **Self-Replication**: RNA can replicate itself, which is a crucial property for the development of life. Certain RNA molecules, known as ribozymes, can catalyze their own synthesis.
2. **Catalytic Functions**: RNA is not just a passive carrier of genetic information; it can also act as a catalyst in biochemical reactions, similar to enzymes made of proteins.
3. **Toolset (pre-genome) Evolution**: RNA acting as "just a passive carrier of genetic information" provides access to a toolset of RNA-based tool encodings.
4. **Prebiotic Chemistry**: The hypothesis posits that RNA could have formed under prebiotic conditions, leading to the emergence of self-replicating systems.
5. **Early Evolution of Life**: The RNA World is thought to be a transitional phase in the evolution of life, leading to the

development of DNA and proteins, which are more stable and efficient for genetic storage and catalysis, respectively.

PAH World Hypothesis

The PAH (Polycyclic Aromatic Hydrocarbon) World hypothesis posits that polycyclic aromatic hydrocarbons, which are complex organic molecules composed of multiple fused aromatic rings, could have played a significant role in the prebiotic chemistry that led to the origin of life [131]. This hypothesis suggests that PAHs, which are abundant in space and can form under various conditions, may have served as scaffolds for the assembly of more complex organic molecules, potentially leading to the emergence of life. PAHs can form through various processes, including combustion, pyrolysis, and in the interstellar medium. Their stability and ability to undergo further chemical reactions make them interesting candidates for prebiotic chemistry. In particular, PAHs can participate in reactions that lead to the formation of more complex organic molecules, such as amino acids and nucleobases. Their planar structure and electron-rich nature allow them to engage in π-π stacking interactions, which could facilitate the assembly of larger molecular structures. The PAH World hypothesis has implications for astrobiology as well, since it suggests that the building blocks of life could be partly synthesized in space and delivered to planets via comets and meteorites, contributing to the prebiotic chemistry on early Earth [132].

Setting the Stage

Like a classic Shakespearean play, let's describe the story of the information encoding molecules of Life in five Acts. The first Act is to set the geophysical stage. The second act will focus on the entry of polymeric nucleic acids in arrangements and concentrations such that we begin to have nucleic information, auto-catalytic self-replication, and RNA World. RNA World may have existed in a pre-cellular phase, with partial compartmentalization or containment (undersea clay beds near volcanic vents, for example), and then been present in the initial proto-cellular expansions into the oceans. In the third Act, clear cell-lines became established according to the two membrane types currently known to exist, and the viral co-evolutionary partners are similarly stabilized into specialization according to these cell lines. Initially, without significant protein-based cytosol processing, a key distinction between cellular life and viral life hasn't evolved yet, thus this phase is sometimes just referred to as Virus World. In the fourth Act, Darwinian selection on cell lines has led to an amino acid (AA) polymer (protein) encoding and production scheme. (AA's available from primitive metabolic and membrane peptide stabilizers.) In the new language of protein sequences 20 common amino acids are coded, plus two rarer amino acids, for a total of 22 amino acids in the protein alphabet. From a maximal information propagation context, it may be that encodings with 22 element alphabets are optimal [4], so this would demonstrate the remarkable power of Darwinian selection to recover this upon optimizing/selecting on protein 'words'. The fifth Act is still happening. It starts with a description of the standard dogma and then describes modifications indicated by recent results that give rise to a much more complex system description. The fifth Act closes with a description of what we must now contend with in terms of systems-complexity karma (such as autonomous, reproducing, AI).

Let's start with a brief description and history of the physical environment to understand the timescales involved. Similarly, a brief description and history is then provided as regards the formation, and 'pre-processing', of the chemical environment. How this might have established the initial conditions for life, and initiated the ensuing phase of Darwinian evolution that brings us to where we are today, is unclear. The sedimentation-ordering emphasis is partly chosen so as to exhibit the pre-processing orderings that may have enabled the proto-biotic possibilities in a broad ranging description of information preserving and propagating contexts. The period of Darwinian evolution eventually settled on the "DNA/RNA→mRNA→Protein & DNA/RNA→ncRNA" models that we see in all living things currently known, as will be discussed. Viruses aren't usually described with this picture, but the genomic data doesn't make sense until viral co-evolution is considered, so the role of viruses will be mentioned throughout.

Beyond gene structure identification, there is the issue of motif identification and reconstructing gene regulatory networks, and that is just the information processing at the DNA level. There is still the mRNA and ncRNA (such as miRNAs) transcriptome, where expression varies according to cell type, cell environment, and cell cycle. For the mRNA's, we can find the proteins that are encoded, for which the bioinformatics task is to then use that information to identify protein, their similarities to other proteins, and their conformation states and function, and with that metabolic pathway identification and function. One of the main sequence analysis tools in RNA and Protein analysis is the same as in DNA analysis, HMMs, although with different emphasis on different issues (alignment instead of internal structure identification, in some cases, for example).

4.1.1 Act 1: Setting the Geophysical Stage

The age of the universe is currently estimated to be: 13.75 ± 0.11 Ga (where 'Ga' is the standard geological term for 'a billion years ago') [133]. Solar Nebula collapse begins around 4.6 Ga, and the Earth accretes from solar nebula around 4.54 Ga. Earth's initial geological formation is mostly done in 10 Million years, with core differentiation and crust formation. Sometime in the 150 Million years after 4.53 Ga the proto-Earth is hit by a large Moon-sized object. Possibly an L5 accretion object ('Theia' hypothesis) that finally oscillates into a collision. Impact thought to be low velocity, grazing, allowing for core exchange (loss) by the smaller body. The result a moon made of crust material and minimal core, and an Earth with a large core (eventually provides a large magnetic 'shield'), a 23.5 axial tilt, and sped-up rotation. (The Moon is the only known object of its size in the solar system that lacks a significant core.) Most theories consider there to be both crust and ocean formation prior to Theia impact. After Theia impact, around 4.4 Ga, the earth's geological ordering processes hit the 'reset' button, with a reset on crust formation, with initial water recovery vaporized by molten surface, e.g., with sterilization of the waters. Eventually, during 4.4 Ga – 4.2 Ga the oceans reform (for the second time). Tectonic plate motion eventually forces up emergent landmasses (the cratons) around 4.0 Ga.(The oldest rocks date to 4.0 Ga, and are found in the cratons, the ocean bottoms are subduction plates, on the other hand, leading to 'memory' being erased every 200 Million years or so). Sedimentation processes and atmospheric erosion begin on the exposed land; have clay formation with water of various levels of salinity, including very little salinity (e.g., fresh-water); and heating of water and clays may be present due to volcanic vent activity. As new land is pushed up by tectonic action, have large collection of water/clay pools and vents, an environment conducive to establishing the Stage 1 of the Chemical environment (described in the next section).

The amount of water found on Earth currently cannot be explained by volcanism and outgassing alone, so must be accounted for partly via cometary impact [134]. Comets are, basically, dirty snowballs. Comets have been found to contain nucleic acids, PAH's, even amino acids (glycine, found on survey by NASA Stardust probe [135]). While not possible to explain the oceans solely from volcanism, it is possible to do so via a reasonable number of cometary impacts (a million over a billion years). Thus, much of our oceans come from comets. The hypothesis of life originating from a sedimentation-ordering is the main thread of this discussion, so life doesn't start with the oceans, but with land formation rising from the oceans, in this hypothesis. Land offers obvious compartmentalization into separate pools, and clay beds, and still has potentially adjacent hydrothermal heating and volcanic venting. Anther leading possibility is of life originating in solvation-ordering in water, possibly originating in underwater hydrothermal vents. If life can directly originate in the primordial oceans, in hydrothermal vents for example, then this opens the possibility for life starting as early as 4.4 Ga on planet Earth. The ocean-vent and clay-bed life origination theories, especially if combined, tend to converge on a RNA World hypothesis, regardless, as will be described in what follows.

If life on Earth is dated to 4.0 Ga, with land emergence, and the earliest archaen fossils date to 3.5 Ga (stromatalites [136] still extant today), that leaves a 500 million year window for RNA World & Virus World dominance. In another 500 million years after the emergence of Archaens, at 3.0 Ga, we have the emergence of oxygenic photosynthesis and the resulting oxygen revolution/catastrophe in the Earth's biome. The critical formation of the chemical environment for life begins with the sedimentation derived clay/organic mixture, possibly heated, with low salinity water, and the chemical environment co-evolves with the virus/cell world from 4.0 Ga to 3.0 Ga, with a shift to a higher oxygen concentration in the atmosphere, among other things.

If life on Earth is dated to the end of the heavy meteorite bombardment phase, at 3.9 Ga, the window for RNA World & Virus World correspondingly shrinks. In recent results the date of earliest archaen fossils has moved from 3.5 Ga up to 3.7 Ga [137]. This would leave a period of only 200 million years for all of the developments indicated for RNA World to occur. Given the latest preliminary results, and the ratchet-like nature of new discoveries closing this 200 million year window even further, it may be that RNA World either 'began in a flash' or happened over a prior 500 Million year period on Mars (with impact-transfer event to get to Earth, see Sec. 4.1.4).

4.1.2 Pre-Life Chemical Environment and the RNA Splice-World Hypothesis

Stage 1. Establishing Initial Chemical Environment – physical ordering. To recap the physical ordering above, we have: (i) sedimentation – microporous clay depositions rich in organic compounds and phosphate [138,139]; (ii) complex solvation – microporous eutectic freezing rich in organic compounds and phosphate; (iii) pooling; and (iv) heating.

Clays, or a surface covering with water in a freeze/thaw cycle, are both ubiquitous upon land formation (tectonic plate action leads to dry land formation, the cratons). Clay minerals are typically formed by erosion of rocks via weathering and small particulate transport to form large beds away from original rock-bound location. Have fresh water (rain) on land, possibly heated by volcanism, with organic components partly from comets (entirely so in case of space "panspermia" [140]). The large aspect of the oceans alone at this stage does not appear sufficient to manage the chemical reactions leading to life without a containment mechanism.

Stage 2. Pre-processing with Chemical Environment: have initial formation of a replicator in the physical-ordering partial containment established in Stage 1. This is 'PAH World', where we have metabolism without genes or full containment. PAHs have been found in a nebula [141], and appear common, giving rise to suggestions of a "PAH World" phase of proto-life [142]. In some origin of life theories it is early metabolism that is the origin of life processes. These theories are often, effectually dual origin theories, or explicitly so, as with Dyson [143], where a nucleic acid information-carrying processes originates separately. In the works of Margulis [144-146] the role of nucleic acid in an infectious role are emphasized throughout (and Dyson references this in his theory for the dual origins, where nucleic-acid enters as a toxic accumulation that turns into an infectious process).

Stage 3. Establishing Initial Replicator: a replicator forms that is partly contained via loose encapsulation inside membrane-forming biomolecules that self-assemble in water due to hydrophobicity properties, or in microporous clays. This is 'Lipid/pre-RNA World' (precursor to Intron World in Sec. 4.2.2), where we have biopolymer pre-genes with minimal metabolism (autocatalytic and self-splicing). Any replicator initiation in a resource-rich semi-contained environment will then propagate to that entire environment: the clay-beds or a eutectic freeze/thaw zone, for example. The dual origin hypothesis mentioned above has a new twist in Sec. 4.1.4, where there's not enough time on early Earth, given the chronology, for the initial nucleic acid information-handling development, so it's possible that this may have been done separately on Mars [127], and then successfully transferred to Earth via impact and meteorite bombardment.

4.1.3 Encapsulation of chemical interactions and information structures

The physical/chemical emergence of proto-biological ordering, from universe formation, to planetary coalescence, up to formation of sediments, with related formation of clay beds (possibly hot with volcanic vent proximity), occurs over a period of around 9.75 billion years (from 13.75 Ga to 4.0 Ga). Earth and Mars [147,148] are examples of sediment-forming planets in our solar system.

Once initiated ca. 4.0 Ga, the biological emergence of fantastically complex information structures occurred over 500 million years via selection. This is due to the rapid emergence of complexity under conditions of selection acting on units of encapsulated, replicative, information structures, in the vast primordial oceans.

Life is known to have occurred at least as early as 3.5 Ga from the fossil record. Viral and primitive single-celled life forms are not viewable in the fossil record, however, so the fossil records don't provide information about precisely the era thought to have occurred when going before 3.5 Ga. So, to be specific about timelines, suppose the 500 million years from 4.0 Ga to fossil record life at 3.5 Ga could be spent in establishing a replicator on Earth (the Mars case to be covered in the next section), and then rapidly disseminating that replicator in the oceans (in a geological-time 'blink') once a self-contained replicator/container system is found -- a fast-start to the fossil record at 3.5 Ga, in other words. For example, consider that from 4.0 Ga to 3.8 Ga Earth establishes a nucleic acid replicator (initiates RNA World, or its PAH +RNA 'artifact-precursor' World); then from 3.8 Ga to 3.7 Ga Earth finds replicator/container (initiates RNA World with cell). From 3.7 Ga to 3.5 Ga Earth has selection and evolution up to the point of life-forms that can be observed in the fossil record (at 3.5 Ga).

So, have biological informatics, or bioinformatics, of encapsulated chemical interactions involving information structures, where those information structures could be RNA-based, or, as will be seen, eventually protein-based, so not just information encoding (the transcriptome), but the information inherent in the protein construct and the related metabolic pathways (the proteome and metabolome).

In the dual origin theories metabolism and nucleic-acid information processing have separate origins, with protein production resulting later (post RNA-World). If asking why Earth would be special to this process it might actually relate to the later phases, particularly the era of protein production and, thus, the standard dogma DNA→mRNA→protein.

4.1.4 Possible Role for Mars [127]

Mars may have had a meteor impact generated transfer of a nucleic acid based replicator system to Earth, providing at least a 500 million year headstart on such development on the geologically 'reset' Earth/Moon system.

Mars cooled approximately 200 million years before Earth. Mars had oceans/sedimentation around 4.5 Ga (while Earth doesn't have oceans until around 4.3Ga).This is partly because Mars is smaller, and thus cooled faster and lower gravity, and partly because Proto-Earth had a developmental reset during this time with its hypothesized L5 'theia' collision that resulted in Earth/Moon system (ca. 4.54 GA): Mars, thus, reached the sedimentation production phase before Earth. Mars had plate tectonics, as revealed by the tell-tale pole-flipping imprint of the magnetic field in the cooling magma of its rift-zone magma fields [147]. Land formation would have been easier on Mars, given its lower gravity. And, the distinctive seasonal layering of sedimentation beds have been conclusively observed on Mars in recent years [148]. (Earth has five beats to the bar, while Mars has the expected ten due to its more complex axial wobble.)

At time of land emergence on Earth (4.0Ga), Mars is hit by a Pluto-sized object (this may be Mars finally crossing paths with one of its own L5 condensates). Mars does not recover much of its oceans and atmosphere…..so where does it go? A small fraction could have made it to earth… and given the inherent replication ability of a nucleic acid replicator that is all that would be needed. If a functional replicator is delivered to Earth, then Earth has a head start by at least 500 million years on replicator development. Since most water-soluble containers (lipid vesicles, etc.) would be easily disrupted or stripped away by such an interplanetary transfer, its more likely that a DNA based replicator in a clay-matrix (possibly pyrite-based) might make such a transfer, to alter earth sedimentation clay beds (or eutectic zones) that have appropriate replicator substrate (such as in the pyrite-based clays of the Rio Tinto Valley [149]).

It is thus conceivable that replicator transfer from Mars Noachian meteor impacts, up to and including the big one (at ~4.0 Ga), would have seeded the Earth's oceans, clay beds, and eutectic zones, etc., upon meteoric transfer to Earth. This would provide up to a 500 million year head start on replicator development. As mentioned in [127], Earth may have cheated on the cradle-of-life exam, and borrowed some answers from Mars, notably a nucleic acid replicator system.

4.2 The original encoding: RNA World

4.2.1 Act 2: Enter Nucleic Acids: Nucleic Information and RNA World

DNA bases are made up of a phosphate group, nitrogenous base (A, G, T, or C), and a (deoxy) ribose sugar (replace T with U for RNA). Single-strand DNA (ssDNA) can anneal with a 'complementary' DNA strand to form a highly stable DNA helix. Annealing is made possible by base-paring between the strands via hydrogen bonds (G-C pair has 3 hydrogen bonds, A-T has 2, G-T has 1). The helix is further stabilized by 'stacking energy' due to the overlapping pi bonds from one planar base-pair in the helix to the next (where the planarbase-paired molecules 'stack' with the next base-pair planar molecule). In RNA or DNA, alternatively, ssDNA can coil onto itself in places where it is G rich, allowing for guanine tetraplexes (G4s) to form (using Hoogsteen hydrogen bonding, not Watson-Crick hydrogen bonding). The G4 motif is $G_{\geq 3}N_xG_{\geq 3}N_xG_{\geq 3}N_xG_{\geq 3}$. A single, planar, G4 (G-quartet) is shown in Fig. 4.1, along with a diagram showing how sacks of G4s give rise to structure (the basis of aptamers both natural and synthetic).

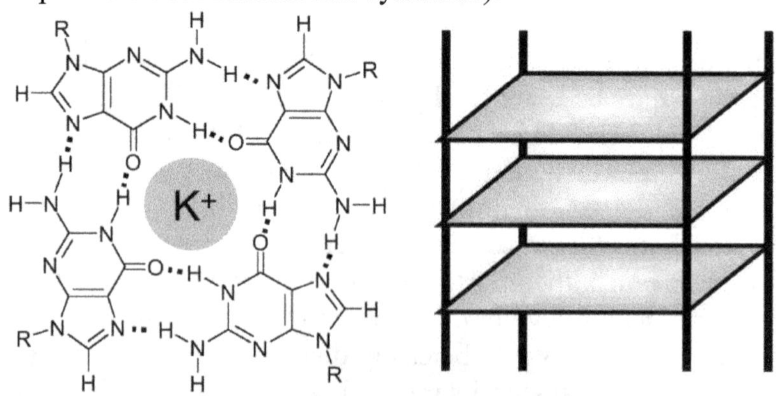

Fig. 4.1. The G4 motif is $G_{\geq 3}N_xG_{\geq 3}N_xG_{\geq 3}N_xG_{\geq 3}$.

Naturally occurring riboswitches (some ribozymes that are still in use), found by Breaker and Nudler in 2002, are nucleic acid based polymers that take on a 3-D conformation with function. This is similar to what is obtained with the aptamers found via SELEX in biotechnology applications.

In order to have ssDNA polymerization, minimally need formation of phosphodiester bonds, where nucleotides are added only at the 3' end. To go from DNA to RNA we need a DNA-dependent RNA polymerase (DdRp) ribozyme. Eventually we arrive at the protein-based DdRp enzyme in use in living organisms today.

RNA Polymerase synthesizes mRNA strands based on DNA template. It was discovered in 1960. Because transcriptional regulation is critical to cells, RNA polymerase is highly regulated. RNA polymerase will bind to promoter region and begin transcription and then release its transcript at specific sites along the DNA sequence. DdRp produces: mRNA; Non-coding RNA (tRNA & rRNA); microRNAs; and Ribozymes. After binding to DNA, RNA polymerase transitions to open conformation and is also able to unwind the DNA about 13 bases upstream. At first RNA polymerase is strongly bound to the promoter. Once longer transcripts begin to be produced, the enzyme clears the promoter. Mg^{2+} ions are used by the enzyme to coordinate phosphate groups during the elongation. In prokaryotes, transcript termination can be Rho dependent/independent. Independent termination involves formation of a hairpin loop in the RNA at the site of termination which helps it to dissociate from the DNA. In eukaryotes RNA polymerase II is responsible for most mRNA production

The transcriptome is the set of all RNAs produced in one or a population of cells. The genome is, roughly, fixed (not strictly fixed due to virus horizontal gene transfer), while the transcriptome can vary with environmental conditions.

Transcriptomics, or Expression Profiling, is the study of mRNA's in a given cell population: for high-throughput studies generally lacking in high accuracy at the single nucleotide level, DNA microarray technology is available. If single-nucleotide level examination of the transcriptome is needed, sequencing technology with cDNAs can be done to obtain a high accuracy result. Nanopore detector based methods may be central, economically, to having both high throughput and high accuracy. (Note the subtlety that the relative amounts of mRNA in a cell are not indicative of the relative amounts of the associated proteins)

4.2.2 Intron World -- a precursor to RNA World

In 'Intron world' we begin with a means for ssRNA extension via a crude RNA-based polymerase, where the polymerase would help to generate, and regenerate, the random pools of ssRNA variants needed. A nucleotide polymerase in the form of an RNA World (pre-cellular) aptazyme is not a strech of the imagination since special cases of such an aptazyme capability still exist in nature, where 1 to 3 bases are polymerized, and some of these cases have been modified to allow almost 100 bases of polymerization [150]. Thus, remnants of an aptazyme polymerization capability are already evident.

RNA/DNA polymers can grow by polymerase extension, aptazyme-based polymer joining, and polymer splicing (via inserts). Notice that in the latter case, splicing can be done without a separate aptazyme or ribozyme, as the splicing replicating element can itself be a ribozyme -- these are known to exist and are known to be ancient (self-splicing intron II and transposons). In this setting, call it the "Intron Earliest" hypothesis (pre-cellular RNA-World), where the first truly replicating element is the intron II / retrotransposon. This allows for an explosion in polymeric complexity at an early stage.

In the Intron World Hypothesis, introns are selected as selfish mobile elements, with self-splicing capability. A more sophisticated Intron spliceosome eventually forms. Introns are eventually demoted to artifacts (or eliminated entirely – in most prokaryotes and viruses), or co-opted as a transcript pause method, or as a spacer for selective modularity (where modularity may not be so much 'functional' as it is a language tokenization with an optimal sequence size according to the specific sequence -- with likely strong dependency on the cumulative hydrophobicity of the amino acids, e.g., a hydrophobicity gematria, for example).

There may have been, thus, intron splicing before RNA→RNA transcriptase capability in Intron World, since it would have more quickly arrived at intron ribozyme (RNA enzyme) self-splicing behavior (Group I + II introns), since selection based on existing RNA self-splicing would have existed first, then would have adopted an artifact to optimize this process and co-evolved on that artifact (a ribozyme before the protein revolution, then a protein-based enzyme afterwards) to achieve improved, faster splicing. The full fledged spliceosome eventually results, variants of which then provide basis for RNA→longer RNA polymerase and eventually RNA→DNA transcriptase (perhaps related to the splice-onto parts of the spliceosome process).

Introns are weird

Some introns are self splicing, and mainly fall into two groups: Group I and II. Group I introns are found in rRNA of lower eukaryotes. The Group I introns have a common secondary structure, and often have catalytic activity for something other than their own self-splicing. Some Group I intron code for endonucleases that promote their own mobility.

In an intron dominated early RNA World (or "Intron World") intron functionalization, and mirtron constructs, could have been the basis for early biochemical information processing.

The adoption of DNA firmly into the RNA World might be marked by the invention of the Group II intron, which allows self-splicing on introns from DNA parent sequence. As with Group I, Group II introns are self-splicing will all sorts extra (beyond self-splicing) catalytic activities. Group II introns, however, use an RNA intermediary during transport to another DNA molecule, at which point they need reverse transcriptase to allow insertion of the RNA back into the new DNA 'host' (if it sounds like a transposon, that's because it does, with different extraction via self-splicing than the standard transposon, otherwise they are very similar and undoubtedly related). Although much more complex (DNA length encodings can be longer than RNA based), Group II introns allowed much more rapid and functionalized developments (via RNA intermediate), that a transition to full "RNA World" could finally be made (with role of DNA as primary information encoding also critically established).

4.2.3 RNA World – (pre-Cellular, initially)

Once you have a lot of random RNA (and DNA) polymers, with self-splicing replicators you've established the framework for autocatalytic RNA polymer complexes with one polymer acting as an information template --- *so the precursor to the modern "gene circuit" could have been in place as far back as the pre-cellular RNA World.*

Suppose RNA World constructs helped launch the first successful 'cell' that could replicate and travel the oceans in their self-replicating cellular compartmentalization. It is likely that proteins would have been beneficial to these early cells if only to help stabilize the cell membrane. Cell-based selection for more and better protein helpers eventually led to the protein-dominated enzymatic systems we see in prokaryotic cells. Given the preceding, the RNA World Hypothesis is as follows.

RNA World Hypothesis (no individualized 'container' yet)

(i) ribozyme replicators: polymerization of nucleotides leads to constructs with self-replicating capabilities (self-replicating, purely RNA-based, ribozymes) [150].

(ii) simple reversible transcriptase between RNA-DNA type ribozyme: introduction of DNA as storage template, and as synthesis template.

(iii) ribosome synthesizer: oligopeptides complex with RNA to form better catalysts, eventually have capability to catalyze peptidyl transfer for primitive protein polymerization. Simple proteins found to be useful, perhaps as bilayer stabilizing and virus interference elements -- eventually arrive at ribosome used in modern protein synthesis.

(iv) protein-based enzymes replace ribozymes (the protein revolution): protein synthesis produces enzymes with more specificity and control and they replace most ribozymes in their various uses.

(v) in Stage 4 to follow, with proto-cell, protein becomes dominant biopolymer, and RNA/DNA take on a predominantly information centric role.

Liposomal 'space program'

What really launched the cell? Some reliable easily reproducible and extendible (for budding) lipid-based membrane material that, basically, self-orders into the desired cell wall purely on the basis of hydrophobicity? Some uniformity would seem to be needed, and indeed, two types of membranes (with very different thicknesses) predominate. Once a reliable cellular unit existed, it could launch from what was probably a 2D chemical environment of a clay-water interface (or even a mostly 1D chemical environment in a channel in a microporous clay). Once a reproducible unit could launch into the 3D ocean habitat, then vastly greater complexity of biomolecular computation could be accessed.

The origin of the prokaryotic cell may be hopelessly obscured by billions of years of time, but the origin of the eukaryotic cell may have some tantalizing clues via the evolution of the spliceosome and RNAi (is discussed later).

4.3 The modern encoding

4.3.1 Act 3: Enter Cell/Virus: Virus World

At the start of the third Act we have compartmentalization, thus cell+virus simultaneously born, giving rise to two forms of transmission of selfish information. Although, the virus seeking compartments, or approximately such, may have existed pre-cellular, according to how things are defined. At this early phase there would be rapid selection and optimization over cell 'compartment' features. The role/benefits of proteins as bilayer constituents may have played a critical role. A tipping point for protein functionality may have been at an early phase in cell membrane evolution to be compatible with protein-based ion-selective channels. It is already hard enough to arrive at an ion-selective channel, bridging the cell membrane, using protein, with its clearly delineated hydrophobic segments. Doing the same with nucleic acid, on the other hand, seems very difficult and likely 'leaky', with likely non-optimal function even if discovered. (It turns out it is possible to engineer nuclei acid based channels that are ion selective [151], known as nucleic acid origami, but the constructs are unlikely to arise in Nature, or if they did, would be very leaky.)

Stage 4. Replicator/container emerges that conquers (populates) the oceans of the world. This is the ancient Virus/Cell World. Lipids and amphiphilic peptide compounds have both hydrophobic and hydrophilic portions. Coupled with mild agitations, such as a wave-zone that is crashing and foaming in a replicator-infested region (if bordering on an infested clay bed or eutectic freeze/thaw zone), then we have what is likely to have been a frequently occurring scenario with conditions that allow replicator containment in lipid vesicles, etc., where primitive budding of the vesicle may have been the basis for reproduction on this first replicator/container arrangement. Eventually some stable replicator/container arrangement will be selected by its successful spread into the oceans.

The dominant replicator/container system may have been a primitive RNA-based system. This may have been the initialization of the primitive cellular RNA World hinted at by the various RNA-based artefacts in all known living organisms. This is 'RNA World with proto-cell', where we have metabolism with primitive genes and individualized containment. In the case of RNA World with container (individualized, cellular) – the containment is itself an important artifact.

Stage 5. Natural selection acts on the stable set of simple RNA World cellular/viral entities. Eventually select on RNA World entity that has better fitness due to co-evolution with artifact, in this case a primitive protein, or protein pre-cursor, and the ability to produce/acquire and manipulate that 'artifact' to added benefit in evolutionary selection. As mentioned, we see no RNA World artifacts involving ion-selective channels (that are very small), indicating the early introduction of such was accomplished via a stable set of complex RNA World entities together with protein synthesis forms providing selection advantage in early cellular life. Protein synthesis may have initially been an artifact that simply provided a membrane strengthening result (with selection), or spliceosomal strengthening artifact or added critical compartment stabilization and functionalization with proten-based channels. Darwinian evolution then acts with Natural selection with DNA/RNA → mRNA → Protein and DNA/RNA → ncRNA systems.

In the Modern Synthesis [152], the transmitted unit of heredity is the gene; the unit of selection is the individual; and the evolving unit is the population. Darwin describes the evolutionary mechanism of natural selection in [153], but not the initial formation of first replicating organism. In neo-Darwinism – evolution is driven by natural selection acting on variation produced by simple mutation and cross-over mutation. In more recent versions of evolutionary theory, Richard Dawkins suggests the gene is the unit of selection [154], as well as non-biological evolutionary systems, such as memes in culture [155], in agreement with prior meme-like ideas described in [156]. More generally, any 'selfish' biomolecule will comprise a unit of selection, and all evolution is co-evolution with such entities. What results is co-evolution with other species (or with virus and other selfish constructs – see next Sec.), or co-evolution with artifact (the Sec. after next).

RNA World type chemical processing also occurs in biotechnology applications: Selection on soups of RNA/DNA molecules is now done in biotechnology in the aptamer selection procedure known as SELEX, where selection isn't for self-replication etc. (as in Nature), but for local binding to some target, with amplification on those found to bind target best.

Role of Viruses and other 'selfish', co-evolving, or parasitic, genomic elements

In the Intron World Hypothesis we have full intron spliceosome before RNA transcriptase replication ability. The premise that the nucleus might be virus-derived (the viral eukaryogenesis hypothesis [157]) is now strengthened in that viral endosymbiogenesis can be accompanied by nucleus formation with a spliceosome at the outset, since it's possible for a virus to have sliceosome machinery given its antiquity. To understand why a virus might contribute its own spliceosome consider the miRNA and RNAi viral defenses of the cell. RNA/DNA silencing methods may date to pre-cellular antiquity. The RNAi defense, on the other hand, may have been largely developed after cell formation, to attack the dsRNA intermediary that is a common molecular complex of most viruses upon introduction to their cellular host (whether originating as ssRNA, or DNA). Thus, there are two mechanisms that might have provided selection pressure for a nuclear membrane: (1) Protection from the RNAi defenses; and (2), separation of translation molecular machinery from pre-processing (splicing) molecular machinery. Splicing by spliceosomes (faster than self-splicing) can be at a rate of one intron every 100-200 seconds (this is organism dependent and highly variable, so just an example). Thus, one role of a nuclear membrane is clear, it must let the slower spliceosome finish its splicing before the translation process can begin.

Thus, the response of some viruses to RNAi may have been to introduce an outer membrane so as to fully construct their single stranded mRNA's and pre-miRNAs (the latter released to cytosol to alter the RNAi-based regulation of gene expression for the cell). The 'nuclear' membrane provided by the virus then allows splicing and dsRNA intermediates to not be exposed to the cell's RNAi defense upon entering the cytosol, and to not have wasteful translation on unspliced pre-mRNA, such that fully constructed ssRNA can be released outside the viral outer membrane (such as with the pox viruses). Even though the spliceosome is thrown in as fortuitous event at this point, such a selection possibility was available, and its amazing success is now apparent, eukaryogenesis -- a virus, with membrane-defense against RNAi, forms the proto-eukaryotic nucleus in an endosymbiotic fusion. It may not be possible to conclusively identify a particular parent in this early genomic sharing, or a direct fusion mechanism of their parent genes, as reshuffling of genes via mobile genetic elements, and immense amounts of time, blurs many of the distinctions provided by a particular initialization.) We shall see, however, that distinctive transcriptome artifacts of this fusion event may still be present, a statistical footprint that is frozen in the eukaryotic genomes since earliest formation.

So viruses may have seeded the formation of the eukaryotic cell itself. What is only beginning to be appreciated, however, is the extent to which viral co-evolution has both benefited and cursed cell-based lifeforms. Benefit: accelerated evolution on population, horizontal gene transfer enrichment, greater fitness/competitiveness; and the future tools of medicine and nanotechnology machinery in general.

Bacteriophages are the most abundant life form. There are more viral genes than genes in cellular compartments. It was thought that viruses might be precursors to cells as far back as 1922 [158]. Haldane poses an early viral stage in evolution in 1928 [159]. An interesting and critical detail of viruses is that they have host specificity. For bacteriophages the host cell is usually a single species of bacterium (a detail used in 'phage typing').

In the mammalian genome [150], approximately 60% of the genome consists of intergenic regions consisting of mobile elements in various stages of decay. Another 30% of the mammalian genome is introns. Introns are thought to allow greater (more modular, safer) modulation via mobile elements (among other things), such as transposons and virus HGT.

In the RNA World, introns may have provided an early mechanism for rate control. In protein world introns provide a means to tokenize into what are called 'exons', that translate to groupings of 50-100 amino acids, whose protein hydrophobicity gematria score represents mostly neutral charge, etc. We could see similar optimized parameters occurring, with the exon-based protein gematria strings, resulting in the observed exon size range (tokenization size). If timing is nuanced for some reason, this can also be managed via use of intron size. Another odd aspect of introns is that they appear to have a "grammar" of their own, but what it encodes is unknown. One possibility is that the intron "grammar" is meant to perform a reset on the transcriptase/polymer uncoiling so that it has no excessive coiling strain when it reaches exonic regions where high fidelity transcription is most important. Although introns are found in everything, including prokaryotes and viruses, they appear to be selected against in these organisms, however, where there is no nucleus to mature the spliced mRNA, and where timing is streamlined for the fastest replication possible.

The origin of viruses has traditionally been placed into three variant theories: (i) viruses were there from the beginning (see RNA World); (ii) the viruses resulted from plasmids or some other selfish mobile genetic element that became parasitic; (iii) the viruses derived from specialized parasitic bacteria (regressive hypothesis). Although all reasonable ideas in their time, and to some extent all describing phenomena that are found to occur in nature. The main thread of evolutionary genomic data appears to support the RNA World case (i). The new understanding of the extensive role of viruses derives from the amazing complexity discerned in just 5000 of the viral genomes sequenced. There are estimated to be hundreds of millions of viral genomes, so more surprises are probably in store.

Transposons are sequences of DNA that can move or transpose themselves (also known as 'jumping genes' [160]). The transposition can either be 'copy and paste' (Class I) or 'cut and paste' (Class II). Class I retrotransposons copy themselves from DNA to RNA by transcription, then from RNA back to DNA by reverse transcription. The reverse transcriptase is often encoded in the transposon itself (similar to retroviruses, such as HIV). Class II DNAtransposons do not involve an RNA intermediate. Transposase makes a staggered cut at target site, cuts out transposon and ligates to target, with DNA polymerase to fill sticky ends and DNA ligase reconnects the backbone. Insertion sites of DNA transposons may be identified by short direct repeats followed by inverted repeats. Some transposons have lost their ability to synthesize reverse transcriptase or transposase, yet continue to be viable because other transposons are still present in the community that do synthesize reverse transcriptase and/or transposase.

The 'P element' transposon family appears to have entered the fruit fly genome in only the past 50 years, having become *globally* disseminated in that time – this provides a hint at the strong HGT pressures from virus (and other mobile element) co-evolution.

In bacteria, transposons can jump from chromosomal DNA to plasmid DNA and back, allowing for the transfer of genes by this manner. This was alluded to in the description of the proto-eubacterial cell, as a means for genome consolidation. This has been observed in the transfer of antibiotic resistance, for example. When the transposable elements lack genes (entirely selfish construct), they are known as insertion sequences.The most common transposon in humans is the ALU insertion sequence. It is approximately 300 bases long and is found copied between 300,000 and 1,000,000 times in the human genome. The Mariner transposon is found in numerous species (found in everything with habitat near the ocean), including human, suggesting distant horizontal transfers might occur as well.

As with viruses, their more independent counterpart, transposons are mutagens. Many transposons (and viruses) contain promoters which drive transcription of their own transposase. More recently, viruses have been found to contain miRNAs, evidently in an attempt to modulate the cells miRNA regulatory system and the antiviral response of the cells innate immunity defenses. The transposons or proviral mutagens with promoters and/or miRNAs can cause altered expression in host genes, a source of disease and evolution. In addition to what is thought to be a primitive form of RNAi (CRISPRi), Bacteria also protect themselves from viruses, and related selfish genomic elements, by producing enzymes, restriction endonucleases, that cut targeted viral DNA/RNA sequences.

4.3.2 Act 4: Enter Amino Acids: Protein Gematria optimal with 22 letters

Mutual Information reveals Codons

Codon structure is revealed in terms of a three-cycle in mutual information (MI) linkages between nucleotides in prokaryotic genome-wide studies (e.g., dinucleotide MI linkages with different base separations). The three-element codon (with a four-element alphabet) has 64 possibilities. Their mapping to amino acid, or 'stop', is shown in Table 4.1.

	U (middle)	**C(middle)**	**A(middle)**	**G(middle)**	
U (5' base)	phe	ser	tyr	cys	**U (3' base)**
U (5' base)	phe	ser	tyr	cys	**C (3' base)**
U (5' base)	leu	ser	stop	stop	**A (3' base)**
U (5' base)	leu	ser	stop	trp	**G (3' base)**
C (5' base)	leu	pro	his	arg	**U (3' base)**
C (5' base)	leu	pro	his	arg	**C (3' base)**
C (5' base)	leu	pro	gln	arg	**A (3' base)**
C (5' base)	leu	pro	gln	arg	**G (3' base)**
A (5' base)	ile	thr	asn	ser	**U (3' base)**
A (5' base)	ile	thr	asn	ser	**C (3' base)**

A (5' base)	ile	thr	lys	arg	**A (3' base)**
A (5' base)	met (start)	thr	lys	arg	**G (3' base)**
G (5' base)	val	ala	asp	gly	**U (3' base)**
G (5' base)	val	ala	asp	gly	**C (3' base)**
G (5' base)	val	ala	glu	gly	**A (3' base)**
G (5' base)	val	ala	glu	gly	**G (3' base)**

Table 4.1. The codon assignment for the 64 codons. Standard amino acid (AA) abbreviations shown for the 20 (main) AA's. The stop codon (UGA) sometimes codes for selenocysteine, a 21st AA. The stop codon (UGA) codes for a 22nd AA, pyrolysine. Thus, the alphabet on AA's appears to encompass 22 letters, but no more. This is a significant number of letters to have in a sequential transmission (propagation) of information, as outlined in mathematical physics analysis of maximal information propagation [1].

The Ribosome: Nature's fantastic nanomachine

The ribosome is a nucleoprotein consisting of two ribosomal RNA (rRNA) subunits: the small subunit reads mRNA and the large subunit catalyzes the linkage of amino acids. After reading the mRNA, the two subunits split apart. Each ribosome contains the A, P, and E sites: A binds aminoacyl- tRNA; P binds peptidyl-tRNA (growing peptide strand); and E binds free tRNA before it exits the ribosome. Since the core catalytic activity is in the large RNA subunit, ribosomes are classified as ribozymes. Their origin is thus likely ancient, going back to the ribozyme dominated RNA World and the earliest introduction of protein production processes.

Ribosomes can be free or membrane-bound. Free ribosomes are present in the cytosol and because it is a reducing environment, proteins with disulfide bonds cannot be made there. Membrane-bound ribosomes are found in the "rough" endoplasmic reticulum, where proteins are usually part of a secretory pathway.

Peptide bond and 22-letter alphabet

So why is stringing together 'words' with an alphabet of 22 letters better than 4 letters? Perhaps even optimal? To explore and capture information in the larger structural space available (the eventual protein structure-function paradigm), it is found that information propagation is optimal with 22 propagation parameters [1], where in the symbolic sequence context described here, this corresponds to an alphabet of 22 letters, or 22 AAs. Why not simply 22 nucleic acids, in that case, and avoid the big shift to an amino acid based formulation? As has long been noted, this comes down to the adoption of the peptide bond as the polymer linker instead of a phosphodiester bond used in nucleic acids, imparting thereby more rigidity, but still retaining critical degrees of freedom in two rotation angles along the backbone at each C-alpha (see Fig. 4.2).

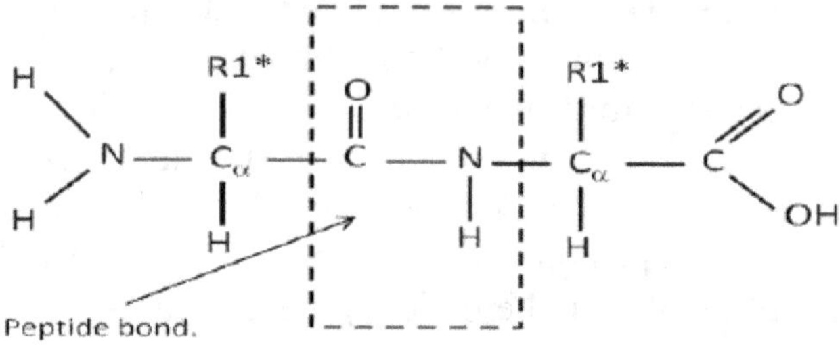

Fig. 4.2. The peptide bond. Shown for the beginning of a homopolymer with R1 residue linking to another R1 residue. The side-chain residue can be nonpolar hydrophobic; polar hydrophilic; or charged (acidic, basic). The peptide bond imparts critical rigidity and leads to alpha helix and other stable conformational structures. A critical bond hybridization leads to planar peptide bond, leaving the polymer's conformational degrees of freedom largely simplified to the two orientation angles along the backbone at each C-alpha.

A protein is a polypeptide biomolecule in stable 3-D conformation. A polypepetide is any length, but implies lacking in complex 3-D conformation. A peptide is a short amino acid oligomer often lacking defined 3-D conformation. After translation, the residues of a protein are often modified by posttranslational modifications (glycosylations, etc.). After release of synthesized polypeptide, the molecule is selected for its ability to form a 3-D conformation with sufficient rigidity and surface attributes to have function.

Protein sequence cannot be modified as easily as nucleic acid sequence. Even a basic splicing operation, which may trace to antiquity in nucleic acids (Intron Earliest Hypothesis) is almost unheard of in amino acid polmers, where proteins with post translational modification by internal sequence splicing are almost unheard of (but do occur, and are known as inteins). This is not to say post-translational modifications are far easier and more significant with proteins in other ways, where elaborate glycosylation structures exist, among other things.

4.4 Dual Origins

Life may have dual origins in informational and metabolic processes

The idea that life may have dual origins in informational and metabolic processes is a concept in the field of astrobiology and the origins of life research. This perspective suggests that life arose not from a single process but rather from the interplay of two fundamental aspects: the informational (genetic) and the metabolic (chemical).

Informational Processes

Informational processes refer to the mechanisms by which genetic information is stored, transmitted, and expressed. The most well-known informational molecule is DNA, which carries the genetic blueprint for living organisms. However, RNA is also considered crucial in the early stages of life due to its ability to both store information and catalyze chemical reactions (as ribozymes). The RNA World Hypothesis posits that early life forms may have relied on RNA for both genetic information and catalytic functions. A seminal paper on this topic is [161]

Metabolic Processes

Metabolic processes involve the chemical reactions that provide energy and building blocks for life. Early metabolic pathways may have been simpler and more primitive than those found in modern organisms. The concept of a "metabolism-first" scenario suggests that life could have originated from self-sustaining chemical networks that eventually led to the development of informational molecules. The Metabolism-First Hypothesis argues that metabolic networks could have formed before the advent of genetic information. A key paper discussing this idea is [162]. The concept of autocatalytic sets, where certain molecules catalyze the production of themselves, is crucial in understanding how metabolic processes could lead to the emergence of life. This is discussed in [26]

Dual Origin Hypothesis

The dual origin hypothesis posits that life emerged from the co-evolution of informational and metabolic processes. This perspective suggests that early life forms may have had simple metabolic networks that eventually led to the development of genetic systems, or vice versa. A paper discussing the interplay between metabolic and informational processes is [130]

Act 5: Enter dogma, Exit karma
Once a DNA → RNA → Protein in a membrane encapsulation was reproducing, a rapid evolutionary exploration began for all sorts of protein function.

4.4.1 Cellular nucleic acid information

DNA is replicated by DNA polymerase (DdDp). DNA Polymerase I was discovered by Arthur Kornberg in 1956 from *E. coli.* DdDp's generally require an existing nucleotide (3' OH) on which to add the new nucleotide . Exonuclease (proofreading) activity is found in some but not all DNA polymerases (DdDp's). DdDp polymerases have processivity along their guide strands, and can add multiple nucleotides per second

RNA is *transcribed* from DNA through RNA polymerase activity (DdRp). The ribosome *translates* the information encoded in the mRNA into protein. As mentioned, DNA in coding regions codes for one of the standard 20 amino acids, and a 21st and 22nd that are coded for in archaic organisms. Genes are transcribed from 5' to 3' ends. The region preceding the 5' end is the *cis* regulatory region, and the region following the 3' end is the *trans* regulatory region.

For viruses, there is RNA → protein (initially, to get viral RdRp, etc.) , then RNA → mRNA → protein (capsid), but now using the virus RdRp. In all, viral genomes can be constructed using a variety of number of nucleic acids (standard DNA or RNA bases, and variants) in any number of forms (ss/ds; circular/linear).

In prokaryotes, genes are often linked under joint regulatory control (i.e., operon structure). Eukaryotes also have various DNA motifs that encode information about translation start sites and modulation of transcription. Methylation of Cs in GC rich sites can turn off gene transcription (known as CpG Islands, shown as GC blocks in Fig. 4.3). CpG islands- regions where cytosine is surrounded by guanine, where the cytosine can be methylated at these locations as a way to turn off gene expression

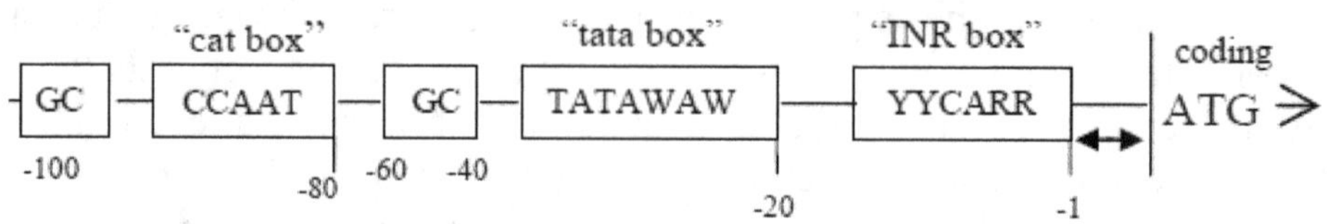

Fig. 4.3. Cis-regulatory control.

C. elegans is odd in that it has some genome structure in prokaryotic style, with operons, and some in eukaryotic style, with introns. Furthermore, *C. elegans* has an 'outron' (Fig. 4.4), essentially an intron on the outside (has a 3' consensus splice site).

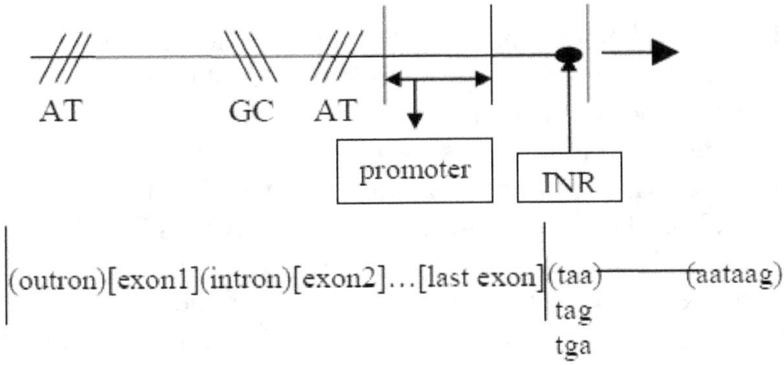

Fig. 4.4. *C. elegens* genome structures.

When introns are cut out, the cut does not necessarily respect the codon boundaries, roughly 50% of the time we get a 'cut' like: (...)(...)(.)---------(..)(...)(...). Synteny, on the other hand, does describe the strict preservation on the *order* of exons, i.e., the specific order of exons that are spliced together to make a mature mRNA is maintained in different species. Introns are significantly more free to change relative to exons, due to more constraint on the latter. The end of gene coding structure helps to produce the poly-A tail found on mature mRNAs (one of many 'tricks' stolen from viruses.)

4.4.2 Cellular system information and co-evolutionary optimization
Transposons are pervasive, just like viruses, making their relation all the more apparent. If looking for a middle-ground between the two, consider the family of endogenous retroviruses (ERVs). As with transposons, the ERVs are thought to endow the host with another mode of competitive viral inhibition against related activities by exogenous retroviruses [163].Transposable elements can be defended by use of siRNAs (small interfering RNAs that trigger the RNAi against the intended target), that perform silencing on the transcribed (mobile) transposon product. In eukaryotes, RNAi is used to inhibit transposon activity in this way (review of RNAi in next section).

Transposition can be classified as either "autonomous" or "non-autonomous" in both Class I and Class II TEs. Autonomous TEs can move by themselves, whereas non-autonomous TEs require the presence of another TE to move. This is often because dependent TEs lack transposase (for Class II) or reverse transcriptase (for Class I).

Co-evolution of virus-human has greatly shaped human molecular evolution and the human genome [164]. Common patterns of nucleotide motif usage are found between virus and host, sometimes moreso than common patterns found in the hosts' nearby species relatives [165].Viruses are hypothesized to play developmental and expression-modification roles. Maternal immune activation (MIA) is found to give rise to pro-viral activation in both mother and fetus. This can lead to a process where the mother gets a viral infection, with the usual inflammatory response, including interleukin-6 (IL-6), in particular. IL-6 alone is found to alter fetal brain development in mice [166], and IL-6, and other stress indicators, also appear to play a critical role in awakening pro-viruses from dormancy [167-169]. The retrovirus XMRV is one such pro-viruses that may be activated during fetal development and is found to have a high correlation to children with autism and adults with schizophrenia. Not all viruses are detrimental to humans, since some viruses target tumor cells (perhaps some ERVs have such benefit). The use of viruses in cancer therapy dates back to the 1940's. In recent advances in this area, in 2007, a virus has been identified that is very efficacious at eliminating neuroendicrine cancers [170].

The evolutionary notion of saltation is evidenced in the viral genomic record via its evident antiquity in some portions, and not in others: evidently selection pressure on the components of an organism can greatly shift, causing the sensitive components of yesteryear to "freeze" in their configurations due to their evolutionary response to the selection pressures becoming minor in comparison to the selection pressures acting on some new component (a specific gene, or antibiotic protein, etc.). Any 'artifact revolution' is hypothesized to be accompanied by a major saltation event in the genomic record. Remarkably, this may allow ancient genomic lineages to still be traceable [150], given the immense geologic time that should have erased these lineages if some portions not 'frozen' [171].

The role of artifact is given in the next section. A new exploration area/method in Bioinformatics, and one that complements efforts in metaheuristics will be described. The new tool involves assessing hypothesized evolutionary mechanism (and linkages) according to metaheuristic optimization performance. The demonstration of the advantages of GA+Viral evolution (both vertical and horizontal gene transfer in host) over standard GA (vertical only), are beginning to be explored [172,173]. For example, if the evolutionary hypothesis to test was regarding artifact revolutions; one could examine introduction of an artifact into a GA-artifact metaheuristic. This is discussed in [127], where GA, GA-artifact, and GA-virus metaheuristics are described and their comparison discussed in the context of testing various evolutionary dynamics (e.g., artifact saltation).

The Cell's RNAi and miRNA/RNAi defenses, and hyper-mutational defenses

In ssRNA gene silencing experiments in 1998 [174] it was found that the ssRNA was not as effective as dsRNA at silencing. In fact, the dsRNA form of silencing resulted in far more silencing than small concentrations could justify, such that a protein catalytic cascade was thought to have been initiated. This was clarified in 2001 [175] with the description of the enzyme Dicer, which cleaves dsRNA into 22bp segments, referred to as small interfering RNAs (siRNAs), that are subsequently inducted into the RNA-induced silencing complex (RISC), which is then able to suppress target sequences similar to the template, here the dsRNA swept up by Dicer. RNA intereference (RNAi) is thus a gene-silencing process, it appears widespread, and thus likely ancient. Even so, RNAi probably only become established after cell formation, after development of the transmembrane TLR3 receptor ('Toll Receptor'), that recognize dsRNA and initiates the interferon response. Once developed, the RNAi machinery provides a highly effected means for viral defense and gene regulation in general.

Depending on the length of dsRNA molecules introduced into the cytosol of a eukaryotic cell there are two responses. For dsRNA longer than 30 base-pairs in length the Toll Receptors are activated, with production of proinflammatory cytokines known as interferons. The interferons, in turn, initate tha adaptive immune response in mammals. For introduction of dsRNA less than 30 base-pairs ('small RNAs') in length a gene-specific response is triggered via the RNAi machinery. There are three types of small RNAs: (1) small interfereing RNAs (siRNAs), between 20-25 bps, typically 21-22 bps, found to be generated by Dicer, an element of the RNAi defense; (2) micro RNAs (miRNAs), generated from genome-encoded hairpin precursors; and (3) repeat-associated small interfering (rasiRNAs), a distinct class of 28 nt siRNAs that are found to silence transposons. Longer dsRNA (29 nt) silencing is found to be stronger at gene silencing than 21 nt templates [176], indicating the significance of the transposon suppression to the fitness of the organism. The small RNAs introduced, or produced by Dicer, are then picked up by the cells RNA-induced silencing complex (RISC).

By introducing dsRNA homologous to target mRNAs those mRNAs can be regulated or silenced (this is done by the genome itself via the miRNAs). Evidently the length of the dsRNA (up to about 29 nts) is a strong indicator of suppression strength by the RISC, as is the degree of match between the miRNA and its intended target. If the intended target is an exact match, or very close, such as occurs with introduction of viral dsRNA, the target is cleaved, and this is enzymatically driven to nearly full gene-silencing. If the intended target is a good match, but with a few mismatches, the RNAi mechanism binds and disables, but does not cut (or very slow cut, so does not go on cutting others as fast), thus partial suppression of expression. The miRNAs operate more in this latter capacity in animals, possibly with a good match to multiple targets, for an overall suppression on a number of genes (and more than one type of miRNA may lead to very controlled gene suppression, allowing further adjustment in extent of suppression and correlation with other, homologous, genes).

The discovery of miRNAs in the encoding of the HIV virus, and other viruses, has led to a new understanding of how viruses have highly evolved strategies for highjacking the cell's RNAi machinery to its own ends [177], including the ability to mediate latent infection ("hide") [178] and modify miRNA expression overall [179]. This has also brought attention to the role of miRNAs in epigentics and human cancer as it is found that the epigenetic machinery of the cell is partly regulated by a subgroup of miRNAs (denoted "epi-miRNAs" in [180]). The role of RNAi and miRNA is only beginning to be understood.

One of the innate defense mechanisms of the cell involves hypermutational biomolecules. HIV researchers have known since 1992 [181] that HIV viruses without the viral infectivity factor (e.g., without the Vif gene), could not infect most human cells. It wasn't until 2002 [182] that this was understood: Vif blocks the DNA-editing enzyme known as APOBEC3G, a cytadine deaminase [183],[184]. If unblocked, the APOBEC3G protein hypermutates the viral DNA to non-functionality. Several other members of the APOBEC family have antiviral activity as well, and the highly radiated structure of the family is highly indicative of strong selection pressure [164]. The APOBEC family are part of a large gene family that includes AID (activation-induced deaminase), the protein responsible for the hypermutation that takes place in the Ig genes during antibody production [185].

4.4.3 RdRp & Viral Eukaryogensis (karma – we are virus)

Pre-eukaryotes may have co-evolved an early form of RNAi with a megavirus, where the battle for RNAi control of host, eventually gave rise to the eukaryotic cell via modulation of RNAi. If nothing else, RdRp's have a pervasive role in eukaryotes in defense against foreign nucleic acids, developmental regulation, genome maintenance, and transcriptome modulation. Viral eukaryogenesis has been proposed in a variety of forms, all consistent with the lack of RdRp in prokaryotes/archaea (as with RNAi), while RdRp is found in eukaryotes and viruses (as is RNAi). So the important role of RdRp for RNAi purposes clearly explains why RdRp would have singular importance to eukaryotes, but is there a non-RNAi role for RdRp as well? In some RdRp RNA productions lengthy antisense strands are produced, and this suggests a possible non-RNAi role, as will be discussed later.

The viral endosymbiosis could have been a very slow transitional process given the likely mutualism 'melting pot' of endosymbionts, including both proteobacteria (eventually leading to mitochondria, etc.) and viruses (including megaviruses) that could have existed at the LECA. This could have existed then since it still exists now, in present day *Acanthamoeba*, which is the host to the megaviruses Mimivirus, Marseillesvirus, and other nucleocytoplasmic large DNA viruses (the NCLDVs), as well as a variety of bacteria, often hosting multiple virus and bacterial visitors at the same time. Furthermore, phylogenetic analysis shows that the NCLDV viruses have an ancient relation to eukaryotes, thus the LECA, placing them in the mutualism context at the time of the hypothesized viral eukaryogenesis. The NCLDV viruses possess their own mRNA capping protein (eIF4E), mRNA capping enzyme, and DdRp, to ensure preferential production of their mRNAs by the host, and eIF4E, at least, is thought to be unique to the virus (not obtained from the host *Acanthamoeba*). The mutualism 'melting pot' environment of the early eukaryote formation is further supported by viral mutualism in the form of favorable RdRp exchange from T7 to the mitochondrial host genome, which is thought to have occurred at the time of the LECA (or earlier) since it is present in all known eukaryotic mitochondria (with only one known exception).

The viral eukaryogenesis theory leaves little opportunity for testing since it relates to a hypothesized ancient historical event, and for this reason testing the viral eukaryogenesis theory has been grouped with other hypothesized endosymbiosis events, such as those endosymbiosis events leading to the mitochondria and chloroplasts. The viral eukaryogenesis hypothesis, however, involves a shift of non-eukaryotic cytosol based genomic data into the newly formed nucleus of the organism, so may leave more evolutionary artifacts in the genomic sequence information than could be obtained in the case of the organelles.

4.4.4 Role of Artifact (karma – we are the tools we use)

We continue to borrow extensively from virus, including the methylated cap and polyA maturation on eukaryotic mRNA (not done by the original prokaryotic mRNA, see Table 4.2 below):

Property	Prokaryotic mRNA	Eukaryotic mRNA
Cis regulation	polycistronic	monocistronic
5' termination	triphosphate	methylated cap (a virus trick)
3' processing	termination to release mRNA	cleavage to release mRNA
3' termination	last base of mRNA	polyA
Transport	n/a (no nuclear membrane)	necessary
Translation	~54 nt/sec	~20nt/sec
Half-life	< 5 min.	> 4 hours
Ribosomes/mRNA	>20	>10

Table. 4.2. Comparison of gene structure and transcript properties for bacterial (single cellular prokaryotic) mRNA and animal (multicellular eukaryotic) mRNA.

The role of artifact in co-evolutionary schemes appears throughout the history of life, and this is only beginning to be understood (see ACO and related methods in [127] for analysis and comparison of such systems in a metaheuristic optimization setting). Some important artifacts: (i) the artifact of containment; (ii) RNA may have been an artifact in PAH World evolutionary selection; (iii) Protein & DNA may have been an artifacts in RNA World evolutionary selection (eventually to become tools outright for more efficient enzymatic activity or more reliable information storage); (iv) Organelles in eukaryotes are thought to be symbiotic co-evolutionary artefacts. The nucleus may be viral in origin (see poxvirus description in viral eukaryogenesis hypothesis), and may have directly led to introduction of sliceosomal introns in the protected nucleus region. The mitochondria still have their own DNA. (v) for human evolution artifacts take on special meaning, as they cover everything from tool use to concept use, such as use of fire; non-verbal concepts; and language. The co-evolution of humans and cultural artifact shows that, not just genes drive evolution, which has long been suspected, but only recently begun to be tested by direct analysis of individual genomic information, such as in recent studies of the serotonin transporter gene [155].

Chapter 5. The Development of Intelligent Life

In this chapter the development of Intelligent Life is discussed. In the previous chapter (and in Appendix B) we addressed the mystery of life itself, and found that much can be understood with modern biology, reducing the 'mystery' to the transition from no life to the earliest replicating life 'process' (PAH World). For intelligent life there is even less mystery as to how it (intelligence) might come about, so in this chapter we simply speak of the 'development' of intelligent life. This is not a long-established sense… as to the reasonableness that intelligent life would typically exist once life exists (with no mystery). With the discovery of neural-net LLM 'capture' at sufficient sophistication (long-range 'attention' learning), however, where rational knowledgeable responses begin to result, we see that if we can get to where there is a neural-net and a learning process, then 'intelligence' might be an expected further outcome. So, to describe the likely development of intelligent life, given the emergence of intelligence once a neural-net with learning is established (as with LLMs described in Chapter 3), we need to see how a neural net might learn *in silico*, in the first place. We've seen that the LLM knowledge capture (scaling) is largely independent of implementation, thus presumably independent of whether *in-silico* or *in vivo*. Thus, even though different structures in the biological realm might arise (e.g., actual brain matter) similar scaling-law information-capture capability, independent of the exact nature of the implementation (biological or not), is considered likely, as long as long-range correlation is sufficiently captured (as described in Ch. 3).

The main effort in this chapter is, thus, to briefly describe statistical learning processes, with focus on the neural-net implementations in AI/ML A short description of statistical learning will be the starting point (Sec. 5.1). Sections 5.2-5.4 then show how neural net learning can be described as motion on a manifold (with minimal relative entropy steps being optimal in many ways). Sections 5.2-5.4 are very mathematical (graduate level mathematical physics or computer science) but can simply be skipped, or skimmed over on math, to get to Section 5.5. To work with neural nets in this setting, a neuro-manifold description is introduced. This requires a review of differential geometry (Sec. 5.2). Once the formalism is clear, the optimal learning process is described very simply, akin to the shortest distance between two points being a straight line in a classical manifold (Sec. 5.3), here we have the quickest learning step is a minimal relative entropy step in the neuro-manifold (Sec. 5.4). Once we see how learning can be done with neural nets, we turn to the application to LLMs with data acquisition that uses Attention to capture long range effects, and uses a Transformer to have sufficient model complexity to exhibit knowledge capture behavior (Sec. 5.5).

5.1 Statistical Learning and Evolving Complexity [186]

Statistical learning involves the use of statistical methods to infer relationships from data. It encompasses a variety of techniques, including regression, classification, clustering, and dimensionality reduction. The goal is to build models that can generalize well to unseen data, allowing for effective predictions and insights. To handle evolving complexity, several statistical learning approaches can be employed:
Regularization Techniques: Methods like Lasso and Ridge regression help manage complexity by penalizing overly complex models, thus preventing overfitting while still allowing for the inclusion of relevant features.
Ensemble Methods: Techniques such as bagging and boosting combine multiple models to improve performance and robustness. These methods can adapt to increasing complexity by leveraging diverse models that capture different aspects of the data.
Online Learning: In scenarios where data arrives sequentially, online learning algorithms update the model incrementally as new data points are observed. This approach is particularly useful in dynamic environments where the data distribution may change over time.
Hierarchical Models: These models allow for the representation of complex structures and relationships within the data. They can adapt to increasing complexity by incorporating multiple levels of abstraction.

5.2 Review of Differential Geometry [187,188]

The following short review draws from texts on differential geometry [187-190].

5.2.1 Differential Topology—Natural Manifold

First recall R^n, the n-dimensional space of vector algebra, with points the n-tuples $(x_1, x_2,\ldots x_n)$ of real numbers. The concept of continuity in R^n is made precise by the study of its topology (where here topology is meant in the local or point-set sense, not the global or algebraic sense).

In the rigorous definition of manifold, to be described later, we find that manifolds have the property that they are "locally R^n". So it is possible that local structures on R^n might carry over to local structures on a manifold. One such structure turns out to be the notion of continuity, and the study of topology provides this and more.

A quick way to demonstrate that a topology is defined on R^n, i.e. that R^n is a topological space, is to introduce the Euclidean distance: d(x,y). One can then define open sets according to unit radius neighborhoods, establish the Hausdorff (continuum) property, etc., and thereby obtain an induced topology from the definition and completeness of the open sets. Topology is more "primitive" then distance, however, non-"distance" measures such as d(p||q) and d(q||p) (the Kullback-Leibler measure) also induce a topology, and it is the same as the Euclidean d(p,q), as can be seen from their local ($p \approx q$) behavior.

Aside from familiarity with topological spaces, R^n being the primary example, an understanding of Manifolds requires an understanding of Mappings (Fig. 5.1). A map from a space M to a space N, $f:M\rightarrow N$, is a rule that associates to an element x of M a *unique* element f(x) of N, $x\rightarrow f(x)$:

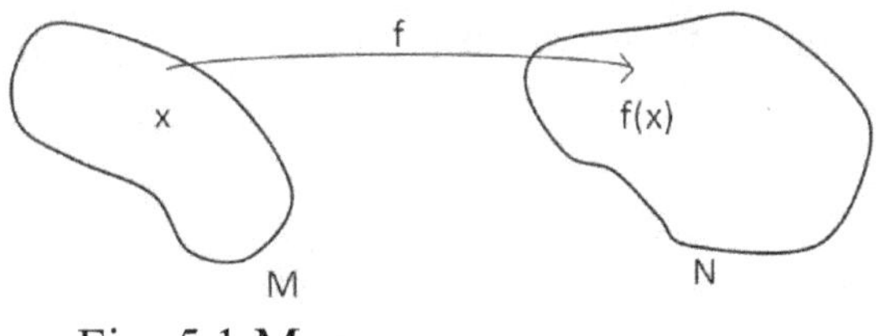

Fig. 5.1 Map.

(a) A map is continuous at x in M if any open set of N containing f(x) contains the image of an open set of M containing X. (Presupposes M and N are topological spaces.) Continuity on all of M is obtained iff the inverse image of every open set of N is open in M.
(b) A map $f:S\rightarrow R^n$ that is continuous is also described as class C^k if all partial derivatives of f(x) of order less than or equal to k exist.
(c) If f is a 1-1 map of an open set M of R^n to N of R^n, then we can define the Jacobian, $J=\partial(f_1,\ldots,f_n)/\partial(x_1,\ldots,x_n)$. If the Jacobian is nonzero at a point, the <u>inverse function theorem</u> assures that the map f is 1-1 and onto in some neighborhood of x.

Definition for Manifold: A set (of "points") M is defined to be a manifold if each point of M has an open neighborhood which has a continuous 1-1 map onto an open set of R^n for some n. ("M is locally like R^n"). Dimension of M is n, local topology is that of R^n.

Since manifolds are "locally R^n" many of the tools of real analysis, defined on R^n, can be used on manifolds also.

Coordinates: by definition, $P \in M \rightarrow (x_1(P), \ldots, x_n(P))$, with $\{x_1, \ldots, x_n\}$ the coordinates of P under the map.

Chart: The pair consisting of a (bijective) neighborhood and its map (Fig. 5.2):

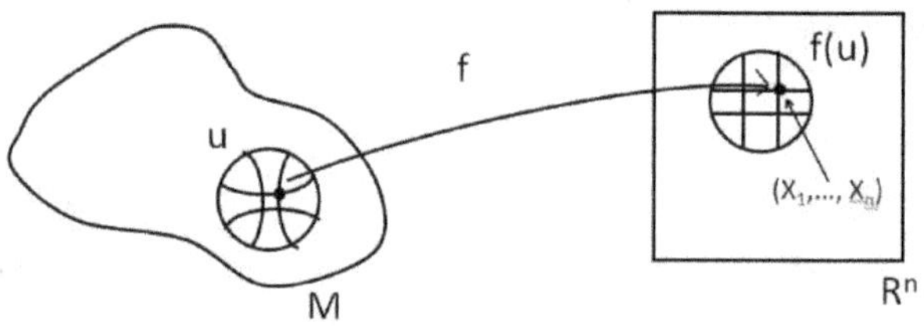

Fig. 5.2 Chart.

Overlapping Charts (coordinate transformations), shown in Fig. 5.3 below:

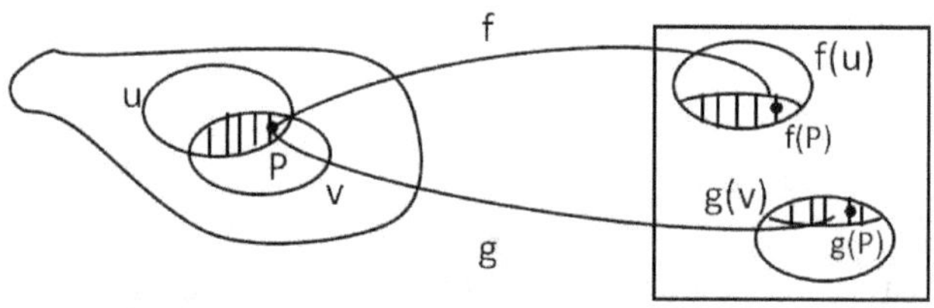

Fig. 5.3 Overlapping Charts.

The coordinates are C^k related if partial derivatives of order k or less are continuous.

Atlas: every point in M is in at least one Chart.

C^k manifold: Manifold with Atlas whose every Chart is C^k related on overlaps.

($\Rightarrow$ A Differentiable Manifold if at least C^1)

Natural Structures on a differentiable manifold:

Curve: a differentiable mapping from an open set of R^1 into M (Fig. 5.4):

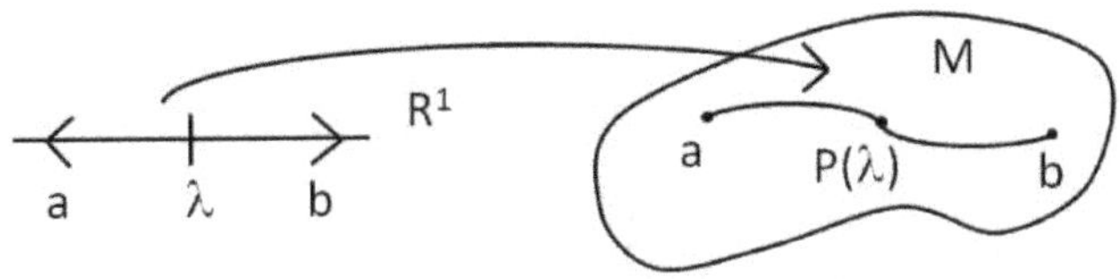

Fig. 5.4 Curve. Note: The parameterization λ defines different curves, even if the image is the same.

Function: a function on M is a rule that assigns a real number to each point of M (Fig. 5.5). If M maps differentiably to R^n, the function induces a function on R^n, which may be differentiable as well.

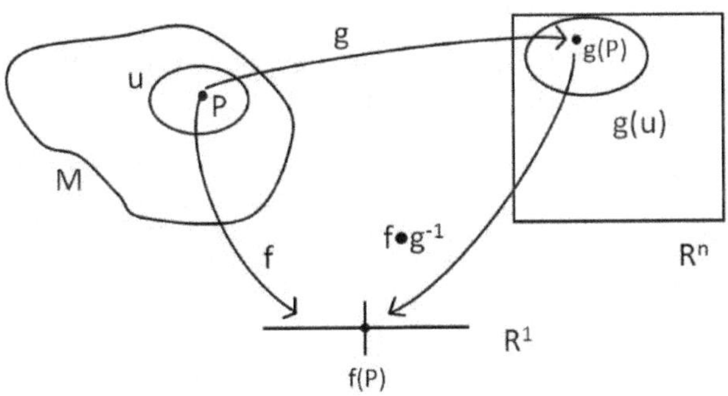

Fig. 5.5 Function.

Vectors and Vector Fields: Consider curves and functions together. This allows one to obtain a differentiable function g (λ) which gives f at a point with parameter value λ : $g(\lambda)=f(x^i(\lambda))$. Now have

(i) The space of all tangent vectors at P and the space of all derivatives along curves at P are in 1-1 correspondence.
(ii) Vectors lie, not in M, but in the tangent space to M at P, called "T_P."
(iii) Vector field, a rule for defining a vector at each point of M.

Basis for Tangent Space T_p: Collection of n linearly independent vectors. Coordinate basis $\{x^i\}$ induces T_p basis $\{\partial/\partial x^i\}$.

Tangent bundle (fiber bundle): Consider manifold M combined with its tangent space T_p, call it TM (Fig. 5.6).

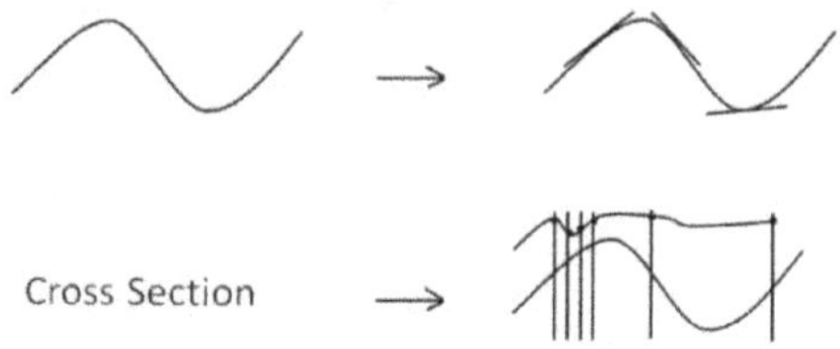

Fig. 5.6 Tangent Bundle, where "cross -section" of TM (gives intuitive notion of "continuous" vector field). Note: even a simple manifold like S^2 has a non-trivial vector bundle

Integral Curves exist for vector fields. Due to uniqueness of solutions the curves never cross except where $v^i=0$. Aside from the $v^i=0$ "caustics," the curves can be manifold filling, and if so you have a congruence (Hamiltonian vector fields).

One-forms: linear real-valued functions of vectors. One-form fields can be described in terms of functions on TM (Fig. 5.7):

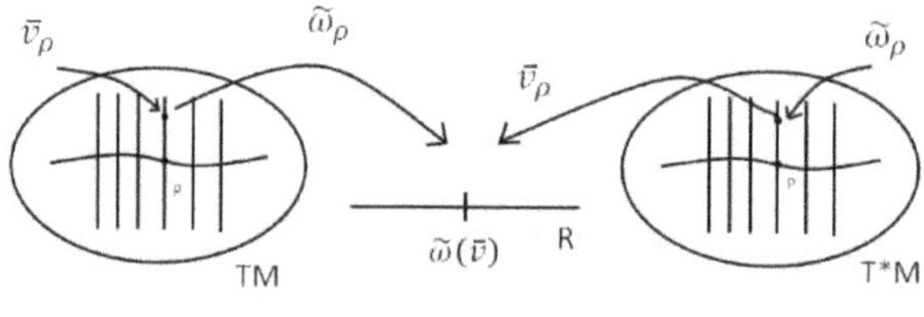

Fig. 5.7 One-forms and cotangent bundle. There is a duality between one-forms and vectors, and the fiber bundle for one-forms, T^*M, is referred to as the cotangent bundle.

Gradient can be interpreted as one-form (Fig. 5.8):

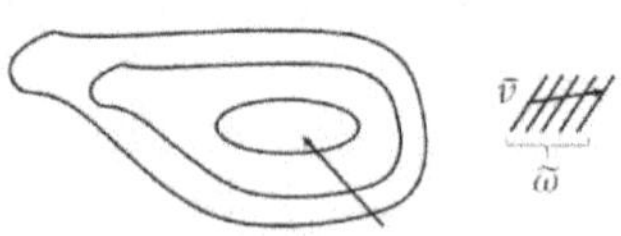

Fig. 5.8 Gradient One-form.

The gradient enables a picture of one-form that is complementary to that of a vector:

Tensor and Tensor Field: A tensor at P is defined to be a linear function which takes as arguments N one-forms and N′ vectors and whose value is a real number.

(i) A scalar is a tensor, such as $v^i w_i$. The v^1 component is not a scalar.
(ii) Metric tensor field: A symmetric tensor field with inverse at every point of the manifold.
(iii) Metric tensors reducible to diag (-1,…,-1,1,…,1)→O(n-k,k) symmetric group.

5.2.2 Differential Geometry –Natural Geometric Structures

The following short review is based on much lengthier analysis at [189].

Consider a tensor $T=T^\beta{}_\alpha e_\beta \otimes w^\alpha$ (where the "$T^\beta{}_\alpha$" are the components and $\{e_\beta\}$is an orthonormal basis in T_p, $\{w^\alpha\}$on orthonormal basis in T_p^*) If we want to compare T at neighboring points in the manifold we must first move the T at one point to the other point. The intuition from vectors in Euclidean space is that we want to move a vector without rotation, i.e., we want to "parallel-transport" the vector. What that implies here is the need for an additional structure—a rule for parallel-transport (and one that generalizes to tensors).

The definition for (covariant) derivative then follows from the rule for parallel-transport (see Fig. 5.9). Consider, for example, the "covariant derivative" $\nabla_u T$ of T along a curve f(λ) whose tangent vector is u=df/dλ :

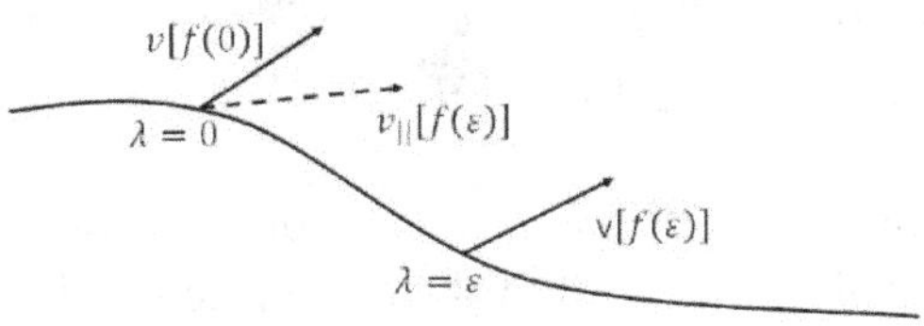

Fig. 5.9 Parallel Transport.

Consider the derivation of the parallel transport as indicated in Fig. 8.9:

$$\vec{u} = \left.\frac{df}{d\lambda}\right|_{\lambda = 0}$$

$$\vec{\nabla}_u T = \lim_{\varepsilon \to 0} \left\{ \frac{T(f(\varepsilon))|\{parallel\ trans.\ to\ f(0)\} - T(f(0))}{\varepsilon} \right\}$$

$$\overline{\nabla} T = \overline{\nabla}(T^{\beta}{}_{\alpha} e_{\beta} \otimes \omega^{\alpha})$$
$$= \overline{\nabla}(T^{\beta}{}_{\alpha}) e_{\beta} \otimes \omega^{\alpha} + T^{\beta}{}_{\alpha}(\overline{\nabla} e_{\beta}) \otimes \omega^{\alpha} + T^{\beta}{}_{\alpha} e_{\beta} \otimes \overline{\nabla} \omega^{\alpha}$$

Introduce standard connection coefficient notation:

$$\overline{\nabla}_{\lambda} e_{\beta} = \Gamma^{\alpha}{}_{\beta\lambda} e_{\alpha}$$

$$\overline{\nabla}_{\lambda} \omega^{\alpha} = -\Gamma^{\alpha}{}_{\beta\lambda} \omega^{\beta}$$

$$\overline{\nabla}_{\lambda} T = (T^{\beta}{}_{\alpha,\lambda} + T^{\mu}{}_{\alpha} \Gamma^{\beta}{}_{\mu\lambda} - T^{\beta}{}_{\mu} \Gamma^{\mu}{}_{\alpha\lambda}) e_{\beta} \otimes \omega^{\alpha}$$

$$T^{\beta}{}_{\alpha;\lambda} = T^{\beta}{}_{\alpha,\lambda} + T^{\mu}{}_{\alpha} \Gamma^{\beta}{}_{\mu\lambda} - T^{\beta}{}_{\mu} \Gamma^{\mu}{}_{\alpha\lambda}$$

Geodesic: A geodesic is a curve that is "straight" and uniformly parameterized as measured in each local R^n Chart. Thus, a geodesic is a curve that parallel-transports its tangent vector u along itself: $\nabla_u u=0$. Simple conceptually, now for concrete implementation. Introduce coordinate system x^{α} and describe u as $u^{\alpha}(0)=dx^{\alpha}(\lambda)/d\lambda|_{\{\lambda=0\}}$:

$$\frac{d^2 x^{\alpha}}{d\lambda^2} + \Gamma^{\alpha}{}_{\mu\lambda} \frac{dx^{\mu}}{d\lambda} \frac{dx^{\lambda}}{d\lambda} = 0$$

Recap of Global Structures [190]

The review in this section largely draws from [190].

n-frame: a set of vector fields $V_{(1)}, V_{(2)}, \ldots, V_{(n)}$ defined on some subset of an n-dimensional manifold M such that they are linearly independent at each point of their domain of definition.

A Chart on $U \subset M$ with coordinates $x^1, x^2, \ldots x^n$ determines an n-frame ∂/∂^x on U. In general there is no n-frame defined on all of M.

Parallelizable Manifold: A global n-frame exists. In general, however, it is difficult to find even a single vector field that is defined globally (hedgehog theorem: a smoothly combed hedgehog has at least one point of baldness→i.e., even dim sphere). Cartesian products of parallelizable manifolds are parallelizable. (In GR one usually has $M=R\times\Sigma_3$, and if a 3 dim. manifold is parallelizable it is orientable. Orientability usually assumed, so parallelizability common in GR).

Connections: The information inherent in the connection coefficients, or in parallel translation, can be associated with principle bundles over a manifold. In general, one considers a bundle space P, over base space M, with structural group G and projection map π: P→ M. It is then possible to establish a one-to-one correspondence between 1 forms on P (Lie algebra valued) and connections on P. These are referred to as connection one- forms.

Affine Connection: If M denotes the Manifold and B(M) its bundle of bases, then a connection on B(M) is called an affine connection.

Difference Forms: Let ϕ and ψ be two connection 1-forms on B(M). Define the difference from $\tau = \psi\text{-}\phi$.

Connections on parallelizable manifolds: Let α be an R^d valued 1-form on M which gives a parallelization of M. Define three connections associated with ρ as follows:

(i) Direct connection: the one for which the vector fields $\rho^{-1}(x)$, for a fixed $x = R^d$, are parallel along every curve.
(ii) Torsion zero connection: the one with the same geodesic as the direct connection but with torsion zero
(iii) Opposite connection: the one with connection ϕ 2τ where ϕ is the direct connection from and $\phi+\tau$ is the torsion zero connection.

$T_{XY} = [X,Y] - \nabla_X Y + \nabla_Y X$ (torsion)
$R_{XY} = \nabla_{[X,Y]} - \nabla_X \nabla_Y + \nabla_X \nabla_Y$ (failure of ∇ to be a Lie algebra homomorphism)

direct: $T_{XY} = [X,Y]$; $R_{XY} = 0$; $\nabla_X Y = 0$.
torsion zero: $T_{XY} = 0$; $R_{XY} = [[X,Y],Z]/4$; $\nabla_X Y = [X,Y]/2$
opposite: $T_{XY} = -[X,Y]$; $R_{XY} = 0$; $\nabla_X Y = [X,Y]$

Let θ be a 2-form and X,Y,Z, vector fields, then:

$$(\nabla_X\theta)(Y,Z) = X\theta(Y,Z) - \theta(\nabla_X Y,Z) - \theta(Y, \nabla_X Z).$$

For metric connection:

$$X\langle Y,Z\rangle = \langle \nabla_X Y,Z\rangle + \langle Y, \nabla_X Z\rangle.$$

For (g, ∇, ∇*), with ∇, ∇* dual connections:

$$X\langle Y,Z\rangle = \langle \nabla_X Y,Z\rangle + \langle Y, \nabla^*_X Z\rangle.$$

With the above concepts reviewed, let's now proceed with the arguments of Amari.

5.3 Amari's Dually Flat Formulation [191-193]

The application of differential geometry methods to the study of statistical models traces back to C.R. Rao in 1945. Rao noted that families of probability distributions could be described by a manifold and that the Fisher information matrix might be taken as a metric on that manifold.

Probability Distributions:

$$p(x) \geq 0 \;\; \forall x \in \chi \text{ and } \Sigma_{\{x\in\chi\}}\, p(x) = 1.$$

Family of Probability Distributions:

$$S = \{p_\theta = p(x;\ \theta) \mid \theta = [\theta^1, \ldots, \theta^n] \in \Phi\}$$

θ parametrizes a n-dimensional statistical model.

Fisher Information Matrix:

$$G(\theta) = [g_{ij}(\theta)];\, g_{ij}(\theta) = E_\theta[\partial_i l_\theta \partial_j l_\theta],$$

$$\text{where } \partial_i = \frac{\partial}{\partial\theta^i} \text{ and } l_\theta(x) = \log p(x;\theta)$$

Sufficient Statistic:

Consider

$$p(x;\theta) = p(x|y;\theta)q(y;\theta).$$

Suppose p(x;θ) = p(x|y;θ) q(y;θ), with associated spaces S≡{p(· ,θ)} and S_F ≡ {p(· ;θ)}, and y is related to x by the constraint y = F(x). If ∀x∈χ p(x|y;θ) does not depend on θ, then F is a "sufficient statistic" w.r.t. S and there results p(x;θ) = p(x|y) q(y;θ). Now, in order to estimate θ, it suffices to know y, and

$$\frac{\partial}{\partial\theta^i}\log p(x;\theta) = \frac{\partial}{\partial\theta^i}\log p(x;\theta) \Rightarrow g_{ij}\ same\ for\ both\ S\ and\ S_F$$

G(θ) is symmetric and positive semi-definite (positive definite with linear independence in $\partial_1 p, \ldots \partial_n p$). From this we can define inner product on the natural basis $[\theta^i]$ and obtain the "Fisher Metric." First revealed in an obscure paper by N.N. Chentsov (1972) [194] it was shown that the Fisher metric could be uniquely singled out based on invariance with respect to "sufficient statistics." *Also invariant w.r.t. sufficient statistics was a class of connections that Amari refers to as α-connections.*

"Invariance" w.r.t. sufficient statistics:

Here the invariance result of Chentsov is restated following the presentation of Amari. In essence, the Fisher metric and α-connections are uniquely characterized by invariance w.r.t. sufficient statistics on arbitrarily reduced probability spaces (reminiscent of Khinchin's reducibility axiom that entropy $H(p_1,p_2,\ldots p_n,0) = H(p_1,p_2,\ldots p_n,)$ in establishing the uniqueness of Shannon's entropy).

The formulation is done for a discrete set and then extended to (continuum) probability densities:

Consider events: $X_n \equiv \{0, 1,\ldots,n\}$; probabilities: $P_n \equiv P(X_n)$; and pairs (g_n,∇_n) of metric and connection on $S \equiv \{P_n\}$. Consider $S_F \equiv \{P_m\}$ where $F: \chi_n \rightarrow \chi_m$ ($n \geq m$, and F is surjective). If F is sufficient w.r.t. S, then g_{ij} and Γ_{ijk} on S and S_F are the same. If sufficiency holds for all n, m, s and F, then g_n is the Fisher metric on P_n and ∇_n is an α-connection on P_n.

Recall Dual Connections:

$$X\langle Y,Z\rangle = \langle\nabla_X Y,Z\rangle + \langle Y, \nabla^*_X Z\rangle.$$

$$\partial_k g_{ij} = \Gamma_{kij} + \Gamma^*_{kji}$$

$$\Gamma^{(\alpha)}_{ijk} = E_\theta\left[(\partial_i\partial_j l_\theta + \frac{1-\alpha}{2}\partial_i l_\theta \partial_j l_\theta)(\partial_k l_\theta)\right], \qquad and\ l_\theta = \log p(x;\theta)$$

Note : The α-connections and (-α) connection are dual w.r.t. the Fisher metric.

The α-connections are defined as follows:

Consider the Exponential Family of distributions in this context:

$$p(x;\theta) = exp\left[c(x) + \sum\theta^i F_i(x) - \psi(\theta)\right]$$

$$\partial_i l = F_i - \partial_j\psi$$

$$\partial_i \partial_j l = -\partial_i \partial_j \psi \longrightarrow no\ x\ dependence$$

for α=1 we then have $\Gamma_{ijk}^{(1)} = -\partial_i \partial_j \psi E_\theta[\partial_k l_\theta] = 0$. So, [$\theta^i$] is a $\nabla^{(1)}$–affine coordinate system, S is $\nabla^{(1)}$–flat. Amari refers to $\nabla^{(1)}$ as the exponential connection, or "e-connection".

Consider the Mixture Family of probability distributions in this context:

$$p(x;\theta) = \Sigma\theta^i p_i + (1 - \Sigma\theta^i)p_0(x)$$

$$\partial_i l = \frac{p_i - p_0}{p};\ \partial_i \partial_j l = -\frac{(p_i - p_0)(p_j - p_0)}{p^2}$$

$$\Rightarrow \partial_i \partial_j l + \partial_i l \partial_j l = 0 \Rightarrow \Gamma_{ijk}^{(-1)} = 0$$

So, [θ^i] is a $\nabla^{(-1)}$–affine coordinate system, S is $\nabla^{(-1)}$–flat. Amari refers to $\nabla^{(-1)}$ as the "mixture connection" or "m-connection".

<u>Divergence</u>: A triplet (g,∇,∇*) can be defined locally from a "divergence" [195], where a divergence D is characterized by:

D(• || •) : S × S → ℜ, where ∀p, ∀q ∈ S × S: D(p||q)≥0, and D(p||q)=0 iff p=q.

<u>Dually flat spaces</u>: If ∇ and ∇* are both symmetric then ∇ flat ↔ ∇*-flat. Since the α connections are symmetric, S is α-flat ↔ S is (-α)-flat, and, in particular, if $\nabla^{(1)}$ is flat then so is $\nabla^{(-1)}$.

If (S, g, ∇, ∇*) is dually flat, then there exists ∇-affine coordinates [θ_i] and ∇*-affine coordinates [η_j]. The inner product between the elements of the tangent spaces is a constant on S, from which it is possible to get the relations:

$$\frac{\partial \eta_i}{\partial \theta^k} = g_{ik}\ \ and\ \ \frac{\partial \theta^i}{\partial \eta_j} = g^{ij}$$

<u>Introduce Potentials</u>: Suppose $\partial_i \psi = \eta_i$, then $\partial_i \partial_j \psi = g_{ij}$, and since g_{ij}, is a metric tensor the partial derivative must describe a positive definite matrix, which in turn implies that ψ is strictly convex. Likewise for $\partial^i \varphi = \theta^i$. In terms of the potentials ψ and φ thus introduced, it is easy to show that the two coordinate systems {θ^i, η_j} can be related by Legendre transformation:

$$\varphi = \theta^i \eta_i - \psi$$

Convexity in potentials that define a Legendre transformation leads to some interesting constructions. If we take our dually flat space (S, g, ∇, ∇^*) with coordinate systems $\{[\theta^i], [\eta_i]\}$ and their potentials $\{\psi, \varphi\}$, then it can be shown that

$$\varphi(q) = \lim_{p\varepsilon S}\{\theta^i(p)\eta_i(q) - \Psi(p)\}$$

and

$$\Psi(q) = \lim_{q\varepsilon S}\{\theta^i(p)\eta_i(q) - \varphi(p)\}$$

Since we have:

$$\Psi(q) - \lim_{q\varepsilon S}\{\theta^i(p)\eta_i(q) - \varphi(p)\} = 0$$

we have:

$$\Psi(q) - \{\theta^i(p)\eta_i(q) - \varphi(p)\} > 0$$

call this D(p||q):

$$D(p//q) = \psi(p) + \varphi(q) - \theta^i(p)\eta_i(q),$$

and it is easily shown that:

$$D(p//q) \geq 0, and\ D(p//p) = 0 \Leftrightarrow p = q$$

Thus, the Exponential and Mixture Families induce a dually-flat space which leads to the fundamental difference measure on distributions being a divergence. The "divergence" form of "distance" function is also indicated in the Link formalism. Although the divergence family is singled out as fundamental at this point, the selection of a specific divergence (like Euclidean distance in the case of metrics) is yet to be determined. In Sec. 5.4 we will see that the "simplest" divergence, the Kullback-Leibler (relative entropy) divergence, is selected when maximizing log likelihood during learning, and that this is, fundamentally, because of the shortest path, or projection theorem, described in Sec. 5.3.2. ***Implicit in this result is that the proper way to measure the difference between distributions is not the Euclidean distance between them but the Kullback-Leibler Divergence between them.***

5.3.1 Generalization of Pythagorean Theorem

Let p, q and r be three points in S. Let γ_1 be the ∇- geodesic connecting p and q, and let γ_2 be the ∇^*-geodesic connecting q and r. If at the intersection q, the curves γ_1, and γ_2 are orthogonal (w.r.t. g), then

$$D(p||r) = D(p||q) + D(q||r)$$

To show this, first consider the γ_1 geodesic:

$$\theta_t^i = t\theta^i(p) + (1-t)\theta^i(q)$$

$$\frac{d}{dt}\theta_t^i\partial_i = \left[\theta^i(p) - \theta^i(q)\right] * \partial_i$$
$$\eta_{ti} = t\eta_i(q) + (1-t)\eta_i(r)$$
$$\frac{d}{dt}\eta_{ti}\partial^i = [\eta_i(q) - \eta_i(r)]\partial^i$$

So, prove relation with:

$$D(p||q) + D(q||r) - D(p||r)$$
$$= \Psi(\rho) + \varphi(q) + \theta^i(p)[\eta_i(r) - \eta_i(q)] - \theta^i(q)\eta_i(r)$$
$$= \left[\theta^i(p) - \theta^i(q)\right][\eta_i(r) - \eta_i(q)] = <$$
$$\left(\frac{d\gamma_1(t)}{dt}\right)_i \left(\frac{d\gamma_2(t)}{dt}\right)^j > = 0$$

5.3.2 Projection Theorem and relation between Divergence and Link Formalism

Suppose M a, submanifold of S, is ∇^*-autoparallel, then $D(p||q) = \min_{\{r\in M\}} D(p||r)$ when the ∇-geodesic connecting p and q is orthogonal to M at q.

Relation to Link formalism, start with:

$$D(p||q) = \psi(p) + \varphi(q) - \theta^i(p)\eta_i(q)$$

Use Legendre transformation:

$$\varphi(p) = \theta^i(q)\eta_i(q) - \psi(q)$$

to get:

$$D(p||q) = \psi(\rho) - \psi(q) + \left(\theta^i(q) - \theta^i(p)\right)\eta_i(q), \text{ where } \eta_i = \frac{d\psi}{d\theta^i}$$

Thus, with shift in notation $\{f=\frac{\partial F}{\partial \omega},\omega\} \rightarrow \{g, \theta\}$ we get the Link formalism used in Ch. 9:

$$D(\omega||\omega') = F(\omega) - F(\omega') - (\omega - \omega')\frac{\partial F}{\partial \omega}\bigg|_{\omega=\omega'}, \text{ where } \frac{\partial D}{\partial \omega} = f(\omega) - f(\omega').$$

The $\{f,\omega\}\leftrightarrow\{g, \theta\}$ duality corresponds precisely to the dually flat connection construction with potentials φ and ψ that are related via Legendre Transformation to the "coordinates" θ and ω. The link function $f(\omega) = \omega$, is associated with square loss and the gradient descent (GD) learning rule. The link function $f(\omega) = \ln(\omega)$, is associated with divergence loss and the exponentiated gradient descent (EG) learning rule. These will be explored in Ch. 9 in the link formalism context, along with an interpolating learning rule between GD and EG given by the link function $f(\omega) = \sinh^{-1}(\omega)$.

5.4 Neuromanifolds [191-193]

The information geometry methods described for families of probability distributions (parametrized by $\{\theta^i\}$) can just as easily be applied to neural networks; where now the parameters are the connection weights (further detailed description of neural nets is in Ch. 9 and Ch. 13). The statistical arguments also carry over and are applicable to stochastic neural nets, i.e., neural networks with noisy input or non-deterministic behavior. As Amari states, "even when a network is deterministic, it is sometimes effective to train it as if it were a stochastic network." [192]. With a stochastic network we then have probability distribution $p(x;\theta)$ and/or conditional probability distribution $p(y \mid x;\theta)$.

Complications associated with repeated observation:

For a single observation we have the distribution $p(x;\theta)$, an element of the family (space) S. For repeated independent observations there is the joint conditional distribution where θ generally changes from one observation to the next. The joint distribution, thus, is an element of a larger family (space)

$$S_T^* = S_1 \times S_2 \times \ldots \times S_T.$$

Multiple observations thus lead to a joint distribution whose manifold dimension (under direct product) increases with the number of observation, while the underlying parameterization is fixed—being the manifold dimension of the family of distributions considered for the individual distribution. This generally leads to a Curved Exponential family description on the joint distribution manifold (where the individual distribution was in an exponential family) ,i.e., a submanifold of the joint distribution manifold is specified. It is possible to describe repeated observations within the framework of the manifold S without referring to the product space S_T^*, but this holds only in the i.i.d. case. In general

$$\theta_T^* = (\theta_1, \theta_2, \ldots, \theta_T),$$

and

$$\theta_T^* = \theta(x_t, u_t),\ t=1,\ldots,T.,$$

where x_t's are given and the u_t's are the only free parameters (the parameters of the underlying neural network). Suppose all x_t are subject to $p(x;\theta)$, $\theta = \theta_1 = \ldots = \theta_T$, where $p(x;\theta)$ is the exponential family $p(x;\theta) = \exp(\theta \cdot x - \psi)$.The joint distribution is

$$p(x_1,\ldots x_T;\theta) = \exp\{\Sigma x_t \cdot \theta - T\psi\}$$

$$p(\bar{x};\theta) = \exp\{\bar{x} \cdot \theta - \psi\}$$

The maximum likelihood estimator (m.l.e.) $\hat{\theta}$ from the observed data $x_1,\ldots,x_T$ is given by maximizing $p(\bar{x};\theta)$:

$$\bar{x} = \frac{\partial}{\partial\theta}\psi(\theta)|_{\theta=\hat{\theta}} = \hat{\eta}$$

The observed data are then represented by the m.l.e $\hat{\eta}$ in S in the η-coordinate system.

Suppose that all x_t are subject to $p(x;\theta)$ where $p(x;\theta)$ is a curved exponential family. Again the observed data $x_1,\ldots, x_t$ are represented by $\bar{x} = \hat{\eta}$, now, however, we have that $\hat{\eta}$ does not necessarily belong to M. The m.l.e. $\hat{u}$, or corresponding distribution $\theta(\hat{u}) \in M$, is given by maximizing the log likelihood $x \cdot \theta(u) - \psi(\theta(u))$ w.r.t. u. Maximizing the log likelihood is equivalent to simply m-projecting $\hat{\theta}$ to M in the Amari differential geometry formalism, and this is equivalent to minimizing the Kullback-Leibler (KL) divergence $K(\hat{\theta} \,||\, \theta(\hat{u}))$ from $\hat{\theta}$ to $\theta(\hat{u}) \in M$. Thus, the KL Divergence is used in the Amari neuromanifold learning process.

EM Algorithm

The following EM/em discussion closely follows Amari [191-193].

Consider $M=\{p(r;\theta(u))\}$ a curved exponential family from which data is regenerated. Data r is observed. Consider $r=r(S_v,S_h)$, a sufficient statistic that includes hidden part S_h. Need to estimate unknown part of r information. Can do this based on observed S_v, and some candidate distribution u′, via the conditional expectation:

E: $$\hat{r}(u') = E[r|s_v;\theta(u')]$$

Now estimate $\log p(\hat{r};\theta(u))$ by conditional expectation also:

E: $$LLH(s_v;\theta(u')) = E[\log p(\hat{r};\theta(u)|\, s_v;\theta(u')] = \theta(u) * \hat{r}(u') - \psi(\theta(u))$$

Now search for better candidate u by maximizing LLH, or equivalently, by minimizing the KL divergence $D(\hat{r}(u')||\theta(u))$ from the guessed data point $\hat{\eta} = \hat{r}(u')$ to μ w.r.t. u.

The algorithm:

Step 0: Initialization step. Guess u_0, the initial guessed distribution $P_0 \in M$ is given by $\theta(u_0)$. Then repeat the following:

Step 1: E-step. Based on candidate probability distribution $P_i \in M$, calculate the conditional expectation of r. This gives the ith candidate for the observed point $Q_i \in D$, whose η-coordinate are $\eta_{(i)}$.

Step2: M-step. Calculate the mle $u_{(i+1)}$ from $Q_i \in D$.

em Algorithm

Search for the pair of points P∈M, Q∈D that minimizes the divergence between D and M, that is (originally proposed in1984 [196]):

$$D(\hat{Q}||\hat{P}) = \lim_{P \in M, Q \in D} D(Q||P)$$

(i) point $\hat{P} \in M$ that minimizes D(Q || P) is given by the m-projection of Q to M (i.e., by the m-geodesic connecting $\hat{P}$ and Q that is orthogonal to M at $\hat{P}$).
(ii) point $\hat{Q} \in D$ that minimizes D(Q || P) is given by the e-projection of P to D (i.e., the e-geodesic connecting P and $\hat{Q}$ that is orthogonal to D at $\hat{Q}$).

The algorithm:

Step 0: Initialization Step. Guess $\hat{u}_0 \Rightarrow \hat{P}_0 \in M$. Then repeat:

Step 1: e-step. e-project $\hat{P}_i$ to D, gives $\hat{Q}_i$.

Step 2: m-step. m-project $\hat{Q}_i$ to M, gives $\hat{P}_{i+1}$.

Relation between EM and em:
(1) M-step and m-step are the same
(2) E-step and e-step differ depending on the conditional expectation of r_h given r_v :

Note: Divergence need not be symmetric or satisfy the triangle inequality, but given "orthogonal" learning steps, a Divergence satisfies a generalized Pythagorean theorem, leading to a the same critical Chapman-Kolmogorov-like propagation rule whether learning step assumes a Euclidean notion of distance (GD) or a KL divergence notion of distance (EG).

Which is Better?
EM is more natural from statistical point of view, but the representation via a neural network does not correspond exactly to statistical inference, therefore em serves better as the approximator ideally suited to the neural network representation of the input-output relation. The two algorithms are asymptotically equivalent when T is large (in the framework of exponential families). If framework extended from exponential family to function space they are exactly equivalent. It may be that a hybrid of GD and EG, or of EM and em type learning, is best.

Amari's Dually Flat formulation is a more natural structure for representing exponential families and neural nets than it might appear at first sight. In particular, the Kullback-Leibler measure is naturally represented in terms of the dual coordinates and their relations via Legendre transformation (with the introduction of appropriate potentials). The potentials, in turn, provide a natural formulation that is precisely that exhibited when using the "link formalism." Similarly, the duality between the coordinates (via Legendre transformation) is precisely that exhibited in the neural net learning algorithms by [197] and [198].

While the e-projection and m-projection are natural geometric notions and certainly a strength of Amari's program, they also represent a weakness. This weakness is perhaps most clearly illustrated in the learning algorithm version where the projection is determined and then the actual update is interpolated (between old data and estimator and new, projected data and estimator). In this instance we must fall back on the gradient descent algorithm, or some such arguments, in order to have stable updates. One approach might be to use the SA link algorithm since it results in a formalism that interpolates between GD and EG learning according to the weight magnitude.

5.5 LLM Learning

5.5.1 Learning with long-range correlations

Learning with long-range correlations refers to the ability of a learning system, such as a neural network, to capture and utilize dependencies that span across long distances in the input data. This is particularly relevant in various domains, including time series analysis, natural language processing, and image processing, where relationships between elements can extend beyond immediate neighbors.

In many real-world datasets, especially sequential data, the relationship between elements can be influenced by events that are far apart in time or space. For example, in natural language, the meaning of a word can depend on context provided by words that are several sentences away [199,200].

Recurrent Neural Networks (RNNs) are designed to handle sequential data and can, in theory, learn long-range dependencies. However, they often struggle with this due to issues like vanishing and exploding gradients. Variants like Long Short-Term Memory (LSTM) networks and Gated Recurrent Units (GRUs) have been developed to better capture these dependencies [201].

Transformers [**14**]: The introduction of transformer architectures has significantly advanced the ability to model long-range correlations. Transformers use self-attention mechanisms that allow them to weigh the importance of all parts of the input sequence, regardless of their distance from each other. This has made them particularly effective in tasks like language modeling and translation.

5.5.2 The self-attention mechanism

The self-attention mechanism is a key component of transformer architectures, which have revolutionized natural language processing (NLP) and other fields. Self-attention allows a model to weigh the importance of different words (or tokens) in a sequence relative to each other, enabling it to capture contextual relationships effectively.

Self-attention computes a representation of a sequence by considering the relationships between all elements in that sequence. It allows the model to focus on relevant parts of the input when producing an output, making it particularly effective for tasks where context is crucial [14, 125, 202].

Here are the Steps involved in the Self-Attention Mechanism:

1. **Input Representation**:

The input to the self-attention mechanism is a sequence of vectors, typically word embeddings. For a sequence of length n, the input can be represented as a matrix X of shape $n \times d$, where d is the dimensionality of the embeddings.

2. **Linear Transformations**:
The input vectors are linearly transformed into three different representations: **Query (Q)**, **Key (K)**, and **Value (V)**. This is done using learned weight matrices: $Q=XWQ, K=XWK, V=XWV$. Here, WQ, WK, and WV are weight matrices of shape $d \times dk$, $d \times dk$, and $d \times dv$ respectively, where dk and dv are the dimensions of the keys and values.

3. **Attention Scores**:
The attention scores are computed by taking the dot product of the query with all keys, followed by scaling: Attention_scores=$dkQKT$. The scaling factor dk is used to prevent the dot products from growing too large, which can lead to gradients that are too small during training.

4. **Softmax Normalization**:
The attention scores are then passed through a softmax function to obtain the attention weights: Attention_weights=softmax($dkQKT$). This results in a matrix where each row sums to 1, representing the relative importance of each key for a given query.

5. **Weighted Sum of Values**:
Finally, the output of the self-attention mechanism is computed as a weighted sum of the value vectors, using the attention weights: Output=Attention_weights$\cdot V$.

Multi-Head Self-Attention

To allow the model to jointly attend to information from different representation subspaces, the self-attention mechanism is often extended to **multi-head self-attention**. This involves:

1. **Multiple Heads**: Instead of having a single set of Q, K, and V matrices, the model has multiple sets (or "heads"). Each head performs the self-attention operation independently.
2. **Concatenation**: The outputs from all heads are concatenated and linearly transformed to produce the final output: MultiHead(Q,K,V)=Concat(head1,…,head$h$)WO where each head is computed as described above.

The paper by Brown et al. [202] discusses the capabilities of the GPT-3 model, which is based on the transformer architecture and utilizes self-attention mechanisms extensively. Here are further self-attention topics and extensions [203, 204]:

1. **Positional Encoding**:

Since the self-attention mechanism does not inherently capture the order of the input sequence (it treats all tokens equally), positional encodings are added to the input embeddings to provide information about the position of each token in the sequence. This is crucial for tasks where the order of words affects meaning.

2. **Computational Efficiency**:

One of the advantages of self-attention is its parallelizability. Unlike recurrent neural networks (RNNs), which process sequences sequentially, self-attention allows for the simultaneous computation of attention scores for all tokens, leading to faster training times on modern hardware.

3. **Scalability**:

Self-attention scales well with the length of the input sequence, but it has a quadratic complexity with respect to the sequence length due to the pairwise interactions between tokens. This has led to research into more efficient variants, such as sparse attention mechanisms and linear attention models, which aim to reduce computational costs.

Variants and Extensions

Techniques like the Reformer [205,206] introduce sparse attention patterns to reduce the computational burden while still capturing long-range dependencies. In tasks like machine translation, cross-attention is used where the queries come from one sequence (e.g., the target language) and the keys and values come from another (e.g., the source language).

5.5.3 Learning with scaling behavior and knowledge capture

The relationship between learning with scaling behavior and knowledge capture is an important area of study in machine learning and artificial intelligence. Scaling behavior refers to how the performance of a learning algorithm changes as the size of the training data or model parameters increases. Knowledge capture, on the other hand, pertains to how effectively a model can learn, retain, and utilize information from the data it is trained on. Here's a detailed exploration of these concepts and their interrelationship:

Learning with Scaling Behavior

Scaling behavior in machine learning often refers to how the performance of a model improves as the amount of training data increases or as the model's capacity (e.g., number of parameters) increases. This can be observed in various contexts, such as supervised learning, unsupervised learning, and reinforcement learning. Research has shown that larger models trained on more data tend to perform better. This phenomenon is particularly evident in deep learning, where models like GPT-3 and BERT demonstrate significant improvements in performance with increased model size and training data. The scaling behavior also relates to generalization, which is the model's ability to perform well on unseen data. As models scale, they often become better at capturing the underlying distribution of the data, leading to improved generalization.

Knowledge Capture

Knowledge capture refers to the ability of a model to learn and retain useful information from the training data. This includes not only memorizing specific examples but also understanding patterns, relationships, and general principles that can be applied to new situations. Knowledge capture can occur through various mechanisms, such as:

- **Parameterization**: The way a model is structured (e.g., neural networks with layers and connections) influences how it captures knowledge.
- **Training Dynamics**: The process of training, including optimization algorithms and regularization techniques, affects how well a model captures knowledge.

Knowledge capture is also evident in transfer learning, where a model trained on one task can be fine-tuned for another task, leveraging the knowledge it has already acquired.

Relationship Between Scaling Behavior and Knowledge Capture

As models scale (in terms of data and parameters), they often exhibit improved knowledge capture. Larger models can learn more complex representations and capture intricate patterns in the data, leading to better performance on a variety of tasks. While scaling can enhance knowledge capture, it also raises concerns about overfitting, especially if the model becomes too complex relative to the amount of training data. Balancing model capacity and data size is crucial for effective knowledge capture.

The interplay between learning with scaling behavior and knowledge capture is a critical area of research in machine learning. As models grow in size and complexity, understanding how they capture and generalize knowledge becomes essential for developing effective and efficient AI systems. By exploring this relationship, researchers can create models that not only perform well on specific tasks but also possess a deeper understanding of the underlying data, leading to more robust and adaptable AI applications. The ongoing exploration of scaling laws, model architectures, and training methodologies will continue to shape the future of machine learning and its applications across various domains.

5.5.4 The Reticular Activating System (RAS) vs the attention mechanism

The Reticular Activating System (RAS) and the attention mechanisms used in AI transformers serve as crucial components in their respective domains—human cognition and artificial intelligence.

Reticular Activating System (RAS)

The RAS is a network of neurons located in the brainstem that plays a key role in regulating wakefulness, arousal, and attention. It filters incoming sensory information and determines which stimuli are worthy of attention. It helps maintain consciousness and alertness, and it is involved in the sleep-wake cycle. The RAS receives input from various sensory pathways and modulates the flow of information to the cortex, effectively prioritizing stimuli based on their relevance or importance. It is influenced by factors such as emotional state, motivation, and the significance of stimuli, allowing for selective attention [207,208].

Attention Mechanisms in AI Transformers

Attention mechanisms in AI, particularly in transformer models, allow the model to weigh the importance of different input tokens when generating an output. This enables the model to focus on relevant parts of the input sequence while processing information. Transformers use self-attention to create contextual representations of words based on their relationships with other words in a sequence. The attention mechanism computes a set of attention scores that determine how much focus to place on each token in the input sequence. This is done through a series of matrix multiplications and softmax functions to normalize the scores. Transformers consist of multiple layers of attention heads, allowing them to capture various aspects of the input data simultaneously.

Comparison

RAS is biological and focuses on regulating attention and arousal in humans, while attention mechanisms in transformers are computational and focus on processing and generating language. RAS filters sensory information based on relevance and emotional significance, whereas transformers use mathematical operations to compute attention scores based on the relationships between input tokens. RAS is influenced by emotional and motivational states, which can affect attention. In contrast, transformers rely on learned representations from large datasets to determine the importance of tokens. The RAS adapts to real-time changes in the environment and individual states, while transformers can be fine-tuned on specific tasks but operate based on fixed learned parameters once trained.

The visual focus system vs the attention system

The visual focus system in humans, which includes mechanisms like critical scanning and saccades, and the attention systems used in AI transformers both serve to prioritize and process information, but they do so in different ways and contexts. Below is a detailed comparison and contrast of these two systems.

Visual Focus System

The visual focus system is responsible for directing visual attention and processing visual information. It involves mechanisms that allow individuals to focus on specific objects or areas in their visual field. Critical scanning refers to the process of examining visual scenes to extract relevant information, while saccades are rapid eye movements that shift the focus of vision from one point to another. These are quick, simultaneous movements of both eyes in the same direction. Saccades allow the visual system to rapidly shift focus between different objects or areas of interest. They can occur voluntarily (to look at something specific) or reflexively (in response to sudden stimuli). This involves the systematic examination of visual scenes, where the eyes may move in a more controlled manner to gather information about the environment. This can include fixations, where the eyes remain still to process visual information, and the duration of these fixations can indicate the importance of the information being processed [209-210].

Chapter 6. The Evolution of Human Communication

In this chapter a description is given of the evolution of human communication. The description begins with evolution of communication in general (not just hominin) where lateralization in the brain (an asymmetry between the hemispheres) is strongly associated with cognitive processes involving communication (Sec. 6.1). Lateralization in modern Homo Sapiens is well documented, and given the evolutionary developments indicated, it is clear that fellow hominins (with whom there was significant interbreeding) also had significant lateralization. Modern Homo Sapiens consist of "Out of Africa" early Homo Sapiens mixing with Neanderthal/Denisovan. These mixing events, in multiple times and places, were not just interbreeding (genetic exchange) but also involved cultural exchange (tools use and other technology transfer). Section 6.2 gives the overlapping hominin timelines and shows significant time for cultural exchange. The Neanderthals/Denisovans will be shown to be of much older lineage (100,000s of years more cultural development) at the time of interbreeding contact, which occurs generally in Northern Europe, or at higher elevations in lower latitudes (Caucasus), and the Siberian Steppe.

Modern humans (homo sapiens) clearly existed 30,000 years ago as evidenced by 'human' cave art and from genetic material. The last known Neanderthals were roughly 40,000 years ago (with their own type of cave art and 'old-style' tools). The use of 'new-style' stone tools begins in the Upper Paleolithic (50,000 years ago) and their use is dominated by the modern humans. A variety of cultural elements other than cave art were also practiced by the Neanderthals, such as burial practices, care for disabled, and these elements, along with the rudiments of language, may have been part of the cultural exchange that would have accompanied interbreeding (in the time 60,000 to 40,000 years ago, at an estimate). The final refinements to human communication, as we know it today, would have occurred in the time from 40,000 years ago until roughly 15,000 years ago. The developments in this time would have not only refined language and music, but also divination practices (or 'games'). In Sec. 6.3 developments in human oral and written communication are described, along with development of music. The description of 'divination games' is detailed in Chapter 7.

6.1 Lateralization

Lateralization refers to the tendency for certain cognitive processes and functions to be more dominant in one hemisphere of the brain than the other. This phenomenon is particularly notable in humans but has also been observed in various other species, including some primates and birds.

Lateralization is thought to confer various adaptive advantages, such as improved efficiency in processing information and enhanced communication abilities. In social species, lateralization may facilitate better coordination and cooperation among individuals. The evolution of lateralized brain functions is believed to be linked to the development of complex behaviors, including language in humans, social interactions in primates, and problem-solving in birds. This specialization allows for more efficient processing of information and may have played a crucial role in the survival and success of these species.

Lateralization is a widespread phenomenon observed in various species, including hominins, primates, birds, rodents, and fish. The degree and nature of lateralization can vary significantly across species, reflecting adaptations to their ecological niches and social structures. In humans, lateralization is most notably associated with language and cognitive functions, while in other species, it may manifest in different behaviors and abilities [211-213].

Birds -- Corvids and Parrots: Some bird species, particularly corvids (e.g., ravens and crows) and parrots, exhibit lateralization in problem-solving and tool use. For instance, studies have shown that these birds may prefer using one foot over the other when manipulating objects, indicating lateralized brain functions [214].

Songbirds -- Lateralization for Song Production: In songbirds, research has shown that the left hemisphere is primarily responsible for the production of learned songs, while the right hemisphere is more involved in the production of innate calls. This lateralization is thought to be linked to the neural circuits involved in vocal learning and communication [215]. Studies have identified specific brain regions, such as the robust nucleus of the arcopallium (RA) and the HVC (high vocal center), that are critical for song learning and production. These areas exhibit lateralized activity during song production, indicating a functional specialization [216].

Rodents: Research on rodents, particularly rats, has shown lateralization in various behaviors, including spatial navigation and problem-solving. For example, studies have indicated that rats may show a preference for using one paw over the other when performing tasks, suggesting underlying brain asymmetries [217]. In cognitive tasks, such as maze navigation, rats have been observed to exhibit lateralized responses, with the left hemisphere being more active during tasks that require complex decision-making [218].

Fish: Some studies have shown that fish exhibit lateralization in their responses to predators and social interactions. For example, certain species of fish may show a preference for using one eye over the other when scanning for threats, indicating lateralized visual processing [219].

Primates (Monkeys and Great Apes) – Lateralization in Tool Use
Studies have indicated that capuchin monkeys show lateralization in tool use, with a preference for using one hand over the other, which may reflect underlying brain asymmetries [220]. Research has shown that chimpanzees exhibit lateralization in tool use and communication. For example, they tend to use their right also hand more frequently for tasks that require precision, suggesting a left-hemisphere dominance for motor control [221].

Primates (Monkeys and Great Apes) – Lateralization in Neuroanatomy
Broca's area is a region in the frontal lobe of the brain that is traditionally associated with language production and processing in humans. Research into the asymmetry of Broca's area in great apes has provided insights into the evolution of language and communication.

In humans, Broca's area is typically found in the left hemisphere of the brain and is associated with language functions. This leftward asymmetry is a key feature of human brain organization. Studies have shown that great apes (such as chimpanzees, bonobos, gorillas, and orangutans) also exhibit some degree of asymmetry in their analog of Broca's area, although the degree and nature of this asymmetry can vary among species. For example, some studies suggest that chimpanzees may show a leftward asymmetry similar to humans, while others indicate that the asymmetry may not be as pronounced.

The presence of asymmetry in Broca's area among great apes suggests that these species may have some capacity for complex communication, although it is not as developed as in humans. Great apes use a variety of vocalizations, gestures, and facial expressions to communicate, indicating a level of social complexity. Research using neuroimaging and post-mortem analyses has shown that the size and structure of Broca's area can differ between species. For instance, a study by Rilling et al. (2008) found that the left hemisphere of the frontal lobe, including areas analogous to Broca's area, is larger in chimpanzees than in gorillas, suggesting a potential link to their more complex social behaviors [222-224].

Prior to Rilling, Cantalupo and Hopkins investigated the size of the Broca's area homolog in chimpanzees and found evidence of leftward asymmetry, similar to what is observed in humans [225]. Asymmetric Broca's areas in the great apes suggests complexity in communication and social interaction as with humans, but not as good as humans. Likewise, we can infer similar degrees of lateralization in hominins (now extinct), and with even greater brain size in the case of the Neanderthals.

With their greater brain size and a similar lateralization, wouldn't the Neanderthal have been smarter, more capable than a Homo Sapiens? Given that the planet is only inhabited by Homo sapiens, the assumption is that the Homo Sapiens were smarter, and a clear transition to better tools is seen with the entry of Homo Sapiens into the archeological record, but might this improvement have occurred in similar timeframe by Neanderthals had they lived? Only

Three key details are known with high confidence: (1) the modern human is 97% anatomical Homo Sapiens ("Out-of-Africa") and 3% Neanderthal (or Denisovan); (2) the Y-chromosomes are exclusively Homo Sapiens ("Out-of-Africa"); and (3) the mtDNA is exclusively Homo Sapiens ("Out-of-Africa"). Something as simple as a fetus-lethal blood-type mismatch would explain why there are no Neanderthal fathers in the modern lineage (this can still happen between modern humans with an Rh factor mismatch between parents). Thus, the 97/3 mix might simply be the result of blood-type compatibility favoring Homo Sapiens in the mixing (further details in a later). Details (2) and (3) could also simply be explained by the larger Homo Sapiens populations simply dominating the genomic material indicated. One result that is firm is the 97/3 mix iself that has clearly occurred (and the 97% Homo sapiens is defined by those existing at that time that may have already undergone some intermixing so not even be 97%). The significance of this is lost if the thousands of years of interbreeding isn't accounted for. What resulted was a gene pool for modern humans that was enlarged with inclusion of Neanderthal/Denisovan genes. From this gene pool was selected the notable immune system genes, but what if another critical selection occurred for further lateralization? We can't see these changes as easily in the genomic record, so there may already have been mixing in the "out-of-Africa" Homo Sapiens population in ways we can't see, or such changes may have come later in the Homo sapiens – Neanderthal interbreeding that only occurred at the end of the interaction period (unlikely). So, the discussions of interbreeding offering better immunity misses the most important element that might have been modified by the interbreeding – greater brain lateralization and improved language ability. Some might argue that the Neanderthal brain was clearly inferior – the "Out-of-Africa" Homo Sapiens came to them (in Europe and the Steppe), adopted then improved their tools, etc. And I agree this is most likely, but it is then argued that the interbreeding could offer no net improvement. This is wrong. Suppose the neanderthal genes impacting brain lateralization were 85% 'bad' and 15% 'good' (in that they offered improved lateralization and linguistic ability). Then selection would simply favor the neanderthal genes that offered better function (over time), and the result is a modern human that is 97/3. Further genomic studies on ancient hominins will clarify this important time in human (genomic) history.

Let's recap the lateralization (Broca's area asymmetry) in hominins: Early hominins, such as *Australopithecus* and *Homo habilis*, had smaller brains compared to later species. Their brain structures were less complex, with limited lateralization similar to that of the primates indicated above. While early hominins demonstrated some cognitive abilities, such as tool use and social interaction, their brain asymmetry was not as pronounced as in later species. The development of larger, and more lateralized, brains in subsequent hominins was accompanied by increased cognitive capabilities.

Homo erectus exhibited a significant increase in brain size (averaging around 900-1,100 cm^3) compared to earlier hominins. This increase is associated with more complex behaviors, including advanced tool use and social structures. Evidence suggests that *Homo erectus* may have begun to show significant brain lateralization, particularly in areas related to motor control and possibly early forms of communication.

Neanderthals had brain sizes comparable to or slightly larger than modern humans (averaging around 1,200-1,750 cm^3). Their brain structure indicates a high degree of complexity, with evidence of lateralization in areas related to vision and motor skills. Neanderthals demonstrated advanced tool-making skills, symbolic behavior, and social organization, suggesting that their brain asymmetry supported complex cognitive functions. However, the specific lateralization patterns may have differed from those of *Homo sapiens*.

In modern humans, Broca's area is located in the left hemisphere of the brain, specifically in the frontal lobe (Brodmann areas 44 and 45). This region is crucial for language production, speech, and grammatical processing. The left hemisphere is typically larger and more developed than the right in *Homo sapiens*, reflecting a strong lateralization for language functions. This asymmetry is thought to be linked to the unique capacity for complex language that characterizes modern humans. Studies using neuroimaging techniques have shown that Broca's area is more active during language tasks, and its structural differences between hemispheres are associated with the ability to produce and understand language.

6.2 Timeline of Hominins and their Tool and Language Developments

4.2 - 1.9 million years ago: Australopithecus [226]

Key Species: Australopithecus afarensis (e.g., "Lucy")
Tool Use: Evidence of simple tool use, such as stone flakes, but not consistently associated with this genus.
Communication: Likely used basic vocalizations and gestures.

2.7 - 1.2 million years ago: Paranthropus [227]

Key Species: Paranthropus boisei
Tool Use: Evidence of using tools, primarily for processing food, but less sophisticated than later hominins.
Communication: Likely similar to Australopithecus, with basic vocalizations and gestures.

2.4 - 1.4 million years ago: Homo habilis [228]

Key Species: Homo habilis
Communication: Likely had more advanced vocalizations than earlier hominins, possibly beginning to use proto-language.
Tool Use: Known as the "handy man," associated with the Oldowan tool culture (simple stone tools).

Oldowan Tools (circa 2.6 million years ago) [228]
Hominins: Associated primarily with *Homo habilis*.
Description: The Oldowan tool culture consists of simple stone tools, such as flakes and cores, used for cutting and processing food.
Significance: Represents the earliest known stone tool technology, indicating the beginning of systematic tool-making.

1.9 million - 110,000 years ago: Homo erectus [229]
Key Species: Homo erectus
Communication: Likely had a more developed form of language, enabling better social coordination.
Tool Use: Associated with the Acheulean tool culture (hand axes and bifacial tools).

Acheulean Tools (circa 1.76 million years ago) [229]
Hominins: Primarily associated with *Homo erectus*.
Description: The Acheulean tool culture features bifacial hand axes and cleavers, which are more sophisticated than Oldowan tools.
Significance: Represents a significant advancement in tool-making, indicating improved cognitive abilities and planning.

Use of Fire (circa 1 million years ago) [230]
Hominins: Evidence suggests that *Homo erectus* was the first to control fire.
Description: Fire was used for cooking, warmth, protection, and social interaction.
Significance: The control of fire allowed for a significant dietary shift, improved nutrition, and social bonding, which may have facilitated communication and cultural development.

700,000 - 300,000 years ago: Homo heidelbergensis [231]
Key Species: Homo heidelbergensis
Tool Use: Advanced tools, including wooden spears and more sophisticated stone tools.
Communication: Likely had a complex form of communication, possibly including early forms of language.

400,000 - 40,000 years ago: Neanderthals [232]

Key Species: Neanderthals (Homo neanderthalensis)
Tool Use: Associated with the Mousterian tool culture (more sophisticated stone tools).
Communication: Evidence suggests they had a form of language and possibly symbolic communication (e.g., burial practices).

Mousterian Tools (circa 300,000 - 30,000 years ago) [233]

Hominins: Associated with Neanderthals and early *Homo sapiens*.
Description: The Mousterian tool culture includes a variety of flake tools, including scrapers and points, made using the Levallois technique.
Significance: Indicates advanced tool-making skills and possibly the use of tools for specific tasks, such as hunting and processing animal hides.

Clothing (circa 100,000 years ago) [234]

Hominins: Likely developed by *Homo sapiens* and possibly Neanderthals. While direct evidence of Neanderthal clothing is scarce, it is believed that they used animal hides for protection against the cold. The presence of tools that could be used for sewing, such as pointed bone implements, suggests they may have fashioned clothing. Additionally, the discovery of Neanderthal remains with cut marks indicates they may have processed animal skins [235].
Description: Evidence suggests the use of animal hides and plant fibers for clothing, which provided protection from the elements.
Significance: Clothing allowed early humans to inhabit a wider range of environments and contributed to social identity and cultural expression.

Caring for disabled

Neanderthals had conditions such as Down syndrome and they cared for children with such conditions [236] . This suggests that Neanderthals were capable of complex social behaviors, including caring for injured or disabled individuals. Archaeological findings have shown that some Neanderthals lived for years despite significant disabilities, indicating that they were cared for by their community [237].

Burial Practices [238]

Neanderthals are known to have engaged in burial practices, which indicates a level of social complexity and possibly a belief in an afterlife. Some burial sites, such as Shanidar Cave in Iraq, show evidence of intentional burial, with grave goods like flowers and tools placed with the deceased. This suggests a ritualistic aspect to their burials.

300,000 years ago – present: Anatomically Modern Humans [239]
Key Species: Homo sapiens
Tool Use: Developed a wide range of tools, including the Upper Paleolithic tool culture (blades, art tools).
Communication: Fully developed language, symbolic communication (art, music), and complex social structures.

Upper Paleolithic Tools (circa 50,000 - 10,000 years ago) [240]
Hominins: Primarily associated with *Homo sapiens.*
Description: This period saw the development of a wide range of tools, including blades, burins, and tools for artistic expression (e.g., engraving).
Significance: Reflects a significant leap in cognitive abilities, creativity, and social organization, as evidenced by the creation of art and complex tools.

6.3 Mixed hominin communities and evidence of cultural exchange [241-244]

The interactions between Homo sapiens and other hominin species, such as Neanderthals and Denisovans, provide fascinating insights into the complexity of prehistoric communities. Evidence suggests that these groups coexisted, interacted, and even interbred in various regions. Let's look into specific examples.

6.3.1 Evidence of coexistence

Homo sapiens and Neanderthals

Evidence indicates that Homo sapiens and Neanderthals coexisted in Europe and parts of Asia for thousands of years. Archaeological sites, such as those in France (e.g., the Grotte de Niaux) and Spain (e.g., El Sidrón), show overlapping artifacts and remains.There is evidence of cultural exchange, including the sharing of tool-making techniques and possibly symbolic practices. For example, Neanderthals are known to have created art and used personal ornaments, suggesting a level of cultural sophistication. Genetic studies have shown that non-African modern humans carry about 1-2% Neanderthal DNA, indicating interbreeding between the two species. This genetic legacy suggests that there were moments of contact and integration.

Homo sapiens and Denisovans
Denisovans were identified from a finger bone found in Denisova Cave in Siberia, dating to around 40,000 years ago. Genetic analysis revealed that they are a distinct group closely related to Neanderthals. Evidence suggests that Denisovans and Homo sapiens coexisted in Asia. Genetic studies indicate that some modern human populations, particularly those in Melanesia and parts of Southeast Asia, carry Denisovan DNA, suggesting interbreeding occurred. While direct archaeological evidence of Denisovan culture is limited, the artifacts found in Denisova Cave, including sophisticated stone tools and jewelry, suggest they had a complex culture.

Specific Archaeological Sites Indicating Interaction [245,246]
***Denisova Cave**:* This site has yielded evidence of both Denisovans and Homo sapiens, including tools and remains that suggest both groups occupied the cave at different times, with potential overlap.
***Grotte de Niaux**:* In France, this cave contains Neanderthal art and tools, and evidence suggests that modern humans may have visited the same regions, indicating possible cultural exchanges.
***El Sidrón Cave**:* In Spain, remains of Neanderthals have been found alongside evidence of modern human presence, suggesting that these groups may have interacted or shared the same environments.

The original "Out of Africa" Theory was posited before the evidence of hominin interbreeding was established from genomic data. It also had a simple view of the movement out of Africa being in one 'wave' (by Homo sapiens). More recent genomic data provides a clearer picture of the wave of hominins moving out of Africa (both Homo erectus , in multiple 'waves', then multiple waves of homo sapiens later). The modern "Out of Africa" Theory posits that modern Homo sapiens originated in Africa and began migrating out of the continent around 60,000 to 70,000 years ago. This theory is supported by genetic, fossil, and archaeological evidence. Key points include:

1. **Genetic Evidence**: Mitochondrial DNA studies suggest that all modern humans can trace their ancestry back to a common ancestor in Africa. The genetic diversity is highest in African populations, indicating that they are the oldest.

2. **Fossil Evidence**: Fossils of early modern humans found in Africa, such as those from Omo Kibish and Herto, show anatomical features that are characteristic of modern humans.
3. **Admixture with Other Hominins**: As modern humans migrated out of Africa, they encountered and interbred with local hominin populations, particularly Neanderthals in Europe and Denisovans in Asia. Genetic studies have shown that non-African populations carry a small percentage of Neanderthal DNA (about 1-2%) and some populations in Asia and Oceania have Denisovan DNA (up to 5% in some groups).

Siberian Steppe Mixing Theory

What is not clarified with the modern Out of Africa theory is when the modern, mixed, human came into existence and in addition to the genetic exchange, what cultural exchange occurred? Use of clothing and tool adoption, at the critical northern mixing area of most long-term cohabitation and exchange, in the Siberian sites (Denisova Cave, etc.), appears to be the best answer available. Thus, what is proposed is an Out of the Steppe Theory for the modern human with modern language capabilities (and divination games, as will be discussed in the next chapter).

6.3.2 No Y-chromosome is from Neanderthal lineage

Neanderthals and anatomically modern humans (Homo sapiens) coexisted in Europe and parts of Asia for thousands of years. Genetic evidence indicates that there was some interbreeding between the two species, which has resulted in modern human populations carrying about 1-2% Neanderthal DNA in their genomes. A study published in *Nature* in 2010 analyzed the genomes of Neanderthals and modern humans and found that while there is evidence of interbreeding, the Y chromosome of modern humans does not show Neanderthal lineage. The study concluded that the Y chromosome in modern humans is derived solely from the Homo sapiens lineage [241]. Subsequent research has reinforced the idea that the Y chromosome in modern humans is not derived from Neanderthals. A study published in *American Journal of Human Genetics* in 2016 analyzed the Y chromosomes of various human populations and found no evidence of Neanderthal contributions [247]. Thus, there is very strong evidence that the Y chromosomes of modern humans do not have Neanderthal lineage. The interbreeding that occurred between Neanderthals and modern humans primarily contributed to autosomal DNA, while the Y chromosome remains distinct to Homo sapiens

6.3.3 Technology Transfer (excluding language)

As *Homo sapiens* migrated into regions inhabited by Neanderthals and Denisovans, they likely encountered their tools and technologies. This overlap in habitats would have provided opportunities for cultural exchange, where modern humans could learn from the established practices of these older hominins. Archaeological findings indicate that some tools used by Neanderthals and early modern humans show similarities, suggesting that *Homo sapiens* may have adopted and adapted techniques from Neanderthal tool-making traditions. For example, the use of the Levallois technique, which was prevalent among Neanderthals, may have influenced the tool-making practices of early modern humans [248].

The emergence of symbolic behavior, such as art and personal ornaments, is a hallmark of the Upper Paleolithic period associated with *Homo sapiens*. However, evidence suggests that Neanderthals also engaged in symbolic practices, such as creating cave art and using pigments. This indicates that cultural practices may have been shared or influenced by interactions between the two groups. Sites like the Chauvet Cave in France and the El Castillo Cave in Spain contain artwork that predates the arrival of modern humans, suggesting that Neanderthals had their own forms of artistic expression. The presence of similar motifs in later Upper Paleolithic art may reflect a continuity of cultural practices or influence [246]

The interbreeding between *Homo sapiens* and Neanderthals has left a genetic legacy in modern non-African populations, with approximately 1-2% of Neanderthal DNA present in their genomes. This genetic exchange may have included advantageous traits related to immune response and adaptation to different environments. And given the above discussion of lateralization, may have led to greater variation in lateralization upon which selection might act to further advantage in language abilities. Similarly, modern humans in parts of Asia and Oceania carry Denisovan DNA, which has been linked to adaptations to high-altitude living, such as in the Tibetan population. This suggests that interbreeding not only facilitated cultural exchange and biological adaptability of modern humans [243], but may have contributed to the initiating of the "Out of Siberian Steppe" movement that would follow.

The chronology of tool use, fire control, and clothing development among hominins illustrates a gradual evolution of technology and culture. The interactions between *Homo sapiens*, Neanderthals, and Denisovans likely facilitated significant technology transfer, enriching the capabilities and cultural practices of modern humans.

6.3.4 Proto-Linguistic Hominin Developments

The study of proto-linguistic hominin ancestors focuses on the evolutionary development of language and communication in early human species. While definitive evidence of language in hominins is challenging to establish, researchers have identified several key traits and developments in our ancestors that likely contributed to the emergence of language.

The evolution of the human vocal tract, including the position of the larynx and the shape of the pharynx, allowed for a greater range of sounds compared to other primates. This anatomical adaptation is believed to have facilitated the production of complex vocalizations. The increase in brain size, particularly in areas associated with language processing (such as Broca's and Wernicke's areas), suggests that early hominins had the neurological capacity for language. The development of the neocortex, which is involved in higher cognitive functions, also played a crucial role [249,250].

The ability to think symbolically is a critical precursor to language. Evidence of symbolic behavior, such as the use of tools and the creation of art, suggests that early hominins were capable of abstract thinking. The development of complex social structures likely necessitated more sophisticated forms of communication. As hominins began to live in larger groups, the need for effective communication to coordinate activities and share information would have increased [251].

Archaeological findings, such as cave art, carved objects, and burial practices, provide insights into the cognitive and communicative capabilities of early hominins. These artifacts suggest a capacity for symbolic thought and social communication. The development and use of tools by early hominins, such as Homo habilis and Homo erectus, indicate advanced cognitive skills and the ability to communicate about complex tasks [252,253].

The concept of proto-language refers to the hypothetical early form of language that may have been used by our hominin ancestors. This form of communication would have been less complex than modern languages but would have included basic vocabulary and grammar structures. Some researchers propose that early hominins used a combination of vocalizations, gestures, and facial expressions to convey meaning, laying the groundwork for the development of fully-fledged languages [254,255]. While the exact nature of these sounds is unknown, it is believed that they were more complex than those of other primates [256].

The social structures of early human groups would have influenced communication methods. Cooperation in hunting and gathering would require effective communication to coordinate efforts and share information about resources [257]

Prehistoric human and hominin communication was likely a multifaceted process that involved vocalizations, gestures, and symbolic representations. These early forms of communication laid the groundwork for the development of complex language and social structures in later human societies [258,259].

6.4 Invention and early use of music in hominins

The invention and early use of music among hominins is a fascinating topic that intersects anthropology, archaeology, and musicology.

6.4.1 Early Evidence of Music

The oldest known musical instruments are flutes made from bird bones and mammoth ivory, discovered in the Swabian Jura region of Germany. These flutes date back to approximately 40,000 years ago, during the Upper Paleolithic period. The discovery of these instruments suggests that early modern humans (Homo sapiens) engaged in musical activities, possibly for social, ritualistic, or communicative purposes [260]. Another significant find is the "Divje Babe Flute," a prehistoric flute made from the femur of a cave bear, found in Slovenia and dated to around 43,000 years ago. This artifact indicates that music-making was a part of the cultural practices of early hominins [261].

The development of music is believed to be closely linked to the evolution of language and social bonding. Music may have served as a means of communication and a way to strengthen social ties within groups. Theories suggest that music could have played a role in group cohesion, emotional expression, and even mate selection [262].

Ethnomusicological studies of contemporary hunter-gatherer societies provide insights into the role of music in early human cultures. For example, among the San people of Southern Africa, music is integral to social rituals, storytelling, and healing practices. These practices likely reflect the functions of music in early hominin societies, suggesting that music was used for communal bonding, spiritual expression, and cultural transmission [263].

6.4.2 Theories on the Origins of Music

Some researchers propose that music may have evolved as a byproduct of other cognitive functions, such as language and social interaction. The "auditory cheesecake" theory, proposed by Steven Pinker, suggests that music is an evolutionary byproduct of the brain's capacity for language and social bonding [264]. Other theories posit that music may have had adaptive functions, such as enhancing group cohesion, facilitating cooperation, and improving communication. The "social bonding" hypothesis suggests that music-making fosters social connections, which would have been advantageous for survival in early human communities [260]. The invention and early use of music among hominins likely emerged as a complex interplay of cognitive, social, and cultural factors [261-263].

6.4.3 Alpha music: 22-states of tonality equivalent to 2x12TET or 24TET [265]

12TET (12-Tone Equal Temperament) and 24TET (24-Tone Equal Temperament) are systems of tuning used in music that divide the octave into equal parts. Starting with a reference frequency of sound, the base note, other notes are related by whole number ratios of that base frequency – these are sounds consonant or pleasing and are the basis of the familiar musical scales. If the notes comprising a complete set of simple whole-number ratio notes are considered, up to doubling of the base note frequency (corresponding to the notes in one octave), then there are 12 such notes. The 12 notes are not perfectly evenly spaced for a given refence, however, and this makes transcriptions between keys impossible, thus the system is slightly modified from the true ratios to nearby frequencies, having equal temperament, thus 12TET.

12TET is the most widely used tuning system in Western music. It divides the octave into 12 equal parts, or semitones. Each semitone is spaced logarithmically, meaning that the frequency ratio between adjacent notes is the same. Specifically, the frequency ratio between two adjacent notes is the twelfth root of 2 (approximately 1.05946). This system allows for the easy modulation between keys and is the foundation for most Western musical scales, including major and minor scales.

24TET divides the octave into 24 equal parts, or quarter tones. Each quarter tone is half the size of a semitone in 12TET, allowing for a greater variety of pitches and microtonal intervals. This system is used in various musical traditions, particularly in Middle Eastern music, and is gaining interest in contemporary Western music for its ability to create more nuanced harmonic and melodic structures.

In 12TET a note not in the reference key, that is used to shift a note in the scale (shifting a semitone sharp or flat), is referred to as a blue note (the basis for the blues scales). Sometimes a predominantly 12TET song will adopt a shift note in the reference scale that is off by a half-sharp or half-flat, e.g. a microtonal shift that would be a 24TET note. Many songs travers two octaves, so, for 12TET, this indicates 24 notes also. A 24-state system of notes seems to be approximately what is in use, or something close to it, which seems tantalizing close to the 22-states suggested from the Alpha numbers in one of their typical presentations.

Ch. 7. The Mystery of Divination: Kabbalah and Tarot

7.1 The Universality of Alpha and the Sacred Numbers

Gematria is based on a 22 letter alphabet. In Hebrew the 22 'letters' are used to label 22 edges that connect 10 nodes (Sephiroth) in a diagram (see Sec. 7.2.2). Other Kabbalistic numbers are encoded in the diagram as well. The original encoding and application (to activities such as divination) for the Jewish Kabbalistic diagram ("Tree of Life") are no longer known. The diagram, however, yields via its edges enumerations: 22, 78=56+22, 137. It yields, by a game-step progression across the diagram, a number of steps comprising 56=7x8 (contiguous path that cycles with permutations), that agrees with a possible progression trough the game-board of Ur (see Sec. 7.4.5.2). The Assyrian Tree, similarly, has an encoding of the key 'kabbalistic' numbers, but also has the number '29' for a full set of Alpha numbers. The Tarot deck, also encodes the kabbalistic numbers. The Tarot originates direct from Babylon/Assyria, as did the Kabbalah diagram and gematria divination and 22-letter 'alphabets' (details in a later section). In Sec. 7.4 the chronology of divination methods will be described and it will become clear that the Chinese divination practices, from the area up to and including the Siberian steppe, are the earliest methods known, and appear to share the same Alpha numbers throughout. This is highly significant, if the "Out of the Siberian Steppe" Hypothesis is correct from Chapter 6. In this chronology we see that early forms of the I-Ching were the starting point (Yarrow Sticks), for which extensions to the standard 64 hexagrams by 14 are possible to 78 (extension by 2 is common), thus directly connecting with the 78 card tarot delineation (actually connecting to a similar 78-state divination system that later maps to the 78-card system familiar to us in modern day, the hypothesized ancient 78-state system is now lost). From Chapter 6, we see how this divinatory 'technology' might have a common origin from the Steppe (up to and including Siberia, where an intersection of H. sapiens, H. Denisovan, and H. Neanderthal is known to have occurred).

7.2 The Gematria Divination Game

Gematria is a word scoring process with refinement. In what follows the rules of gematria will be described, which are based on words formed from a 22-letter alphabet. All ancient scripts are based on 22-letter "alphabets" that derive from Ugaritic. The familiar "abcde…" alphabetical order from elementary school is found in all descendant written languages of the reduced Phoenician Writing System (details to follow), which is all of the ancient world except China. (Gematria is applied to more modern languages, with more than 22 letter alphabets by a reduction to an effective 22-letter alphabet.)

Part of Kabbalistic mysticism involves gematria. Word scores are thought to indicate a hidden power in the word. The best, most 'holy', score for a word is 137. Various forms or attributions of "God" all score 137. The word Kabbalah itself scores as 137. Furthermore, kabbalah is not known to associate with any other meaning in ancient Hebrew. The first documented use of gematria is from an Assyrian inscription dating to the 8th century BCE, commissioned by Sargon II (even though clearly in use long before that). In this inscription, Sargon II states: "the king built the wall of Khorsabad 16,283 cubits long to correspond with the numerical value of his name". Interestingly, much of the Jewish Bible is thought to involve gematria encoding. Think of it like iambic pentameter modifying speech in Shakespearian dialogue, we have the same here, but instead of iambic pentamer, each 'line' must score some gematria value (example from Greek gematria to follow). Text analytics regarding iambic pentameter is given in the next chapter.

Gematria was invented by the Assyrians and Babylonians, or possibly an even earlier group. Thus, there remains a mystery as to the origins of Kabbalah in the form of its gematria roots. Starting with the Assyrians, however, we are able to follow the Kabbalah as it passes to later groups. The tracing of the Hebrew adoption of gematria to the Babylonian captivity, for example, has been made by many sources.

Assyrian gematria

In Assyrian gematria, letters from the Assyrian alphabet (which is a variant of the Aramaic script) are assigned numerical values. This practice allows for the exploration of the numerical significance of words and phrases, often revealing hidden meanings or connections between different concepts. The use of gematria can be seen in various religious and philosophical texts, such as that given above, where scholars and mystics analyze the numerical relationships to gain insights into spiritual truths.

Hebrew gematria

In Hebrew gematria, the letters are assigned values as follows:

- Aleph (א) = 1
- Bet (ב) = 2
- Gimel (ג) = 3
- Dalet (ד) = 4
- He (ה) = 5
- Vav (ו) = 6
- Zayin (ז) = 7
- Chet (ח) = 8
- Tet (ט) = 9
- Yod (י) = 10
- Kaf (כ) = 20
- Lamed (ל) = 30
- Mem (מ) = 40
- Nun (נ) = 50
- Samech (ס) = 60
- Ayin (ע) = 70
- Pe (פ) = 80
- Tsadi (צ) = 90
- Qof (ק) = 100
- Resh (ר) = 200
- Shin (ש) = 300
- Tav (ת) = 400

Gematria is used to find connections between words and concepts by comparing their numerical values. For example, the Hebrew word for "life" (חיים, chaim) has a gematria value of 68, which can be linked to other words or phrases with the same value to uncover deeper meanings.

Other Gematria systems that followed

Since the Assyrians, Babylonians, and ancient Hebrews, were all playing the gematria game its logical to ask if the Greeks did the same. The answer is an emphatic yes. The Greek form of gematria is known as isopsephy. An interesting example of isopsephy is given on a headstone found at the Temple of Artemis at Sparta Orthia. It is a 2nd-century AD example of isopsephic elegiac verse:

> ΟΡΘΕΙΗ ΔΩΡΟΝ ΛΕΟΝΤΕΥΣ ΑΝΕΘΗΚΕ ΒΟΑΓΟΣ ΒΨΛ
> ΜΩΑΝ ΝΙΚΗΣΑΣ ΚΑΙ ΤΑΔΕ ΕΠΑΘΛΑ ΛΑΒΩΝ ΒΨΛ
> ΚΑΙ ΜΕΣΤΕΨΕ ΠΑΤΗΡ ΕΙΣΑΡΙΘΜΟΙΣ ΕΠΕΣΙ ΒΨΛ

which is a votive stele for a boy who won a singing contest. The words in each line have a gematria score of 2730, which is indicated by the "ΒΨΛ" at the end of the line (yes, the gematria scoring was actually carved into the headstone at the end of each line…).

What about the Romans? Did they have a gematria system as well? Yes, so much so that the famous physician Galen [266] was threatened with assassination [267] for not administering the Roman standard of medical care (ca. 162AD) that relied upon divination and mysticism (including gematria). Galen insisted on observation and reason to guide his actions, and was very well learned, and for this reason he was the physician for a succession of Roman emperors. Oddly, Galen's father, Aelius Nicon of Pergamon, was an expert in isopsephy according to Galen [268].

7.2.1 Assyrian Tree-of-Life Encoding: {4, 9, 22, 29, 32, 56, 78, 137}
The earliest form of the tree of life is from the Assyrians (who were clearly doing gematria early given the Sargon quote above). It wasn't just art for art's sake, however, the trees were the same, encoding something. Some have attempted to draw a correspondence between the Assyrian "tree of life" and the Kabbalistic Hebrew "tree of life," but with only limited success. Regardless, there is still the 22 and the 137 basis for the gematria in common, so we might ask how far back can we trace 22 letter systems. For this we must trace to the earliest common "alphabet" based on Cuneiform Ugaritic. Technically the Ugaritic alphabet had 30 symbols but it encoded for four languages, each of which used only 22 letters. To quote Healey: "Clay tablets written in Ugaritic provide the earliest evidence of both the North Semitic and South Semitic orders of the alphabet, which gave rise to the alphabetic orders of the reduced Phoenician Writing System and its descendants (including the Paleo-Hebrew aleph-bet, Hebrew, Syriac, Greek and Latin) on the one hand, and of the Ge'ez alphabet on the other which was also influenced by the ancient Egyptian Hieroglyphic writing system, and adapted for Amharic." [269]

The Assyrian Tree of Life from [270]

Fig. 7.1 The Assyrian Tree of Life.

Figure 7.2 shows just the Tree part:

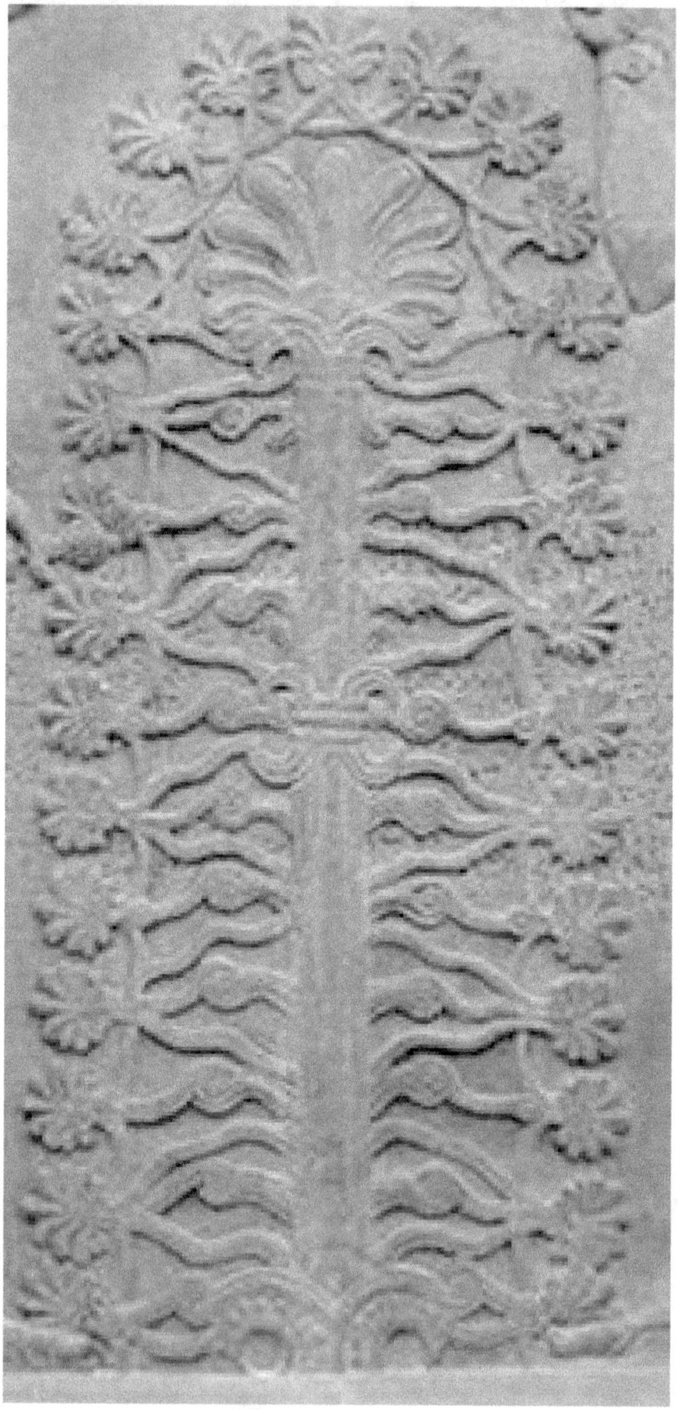

Fig. 7.2 The Assyrian Tree of Life.

Let's count various features in the Tree to see what numerology is encoded:
(1) The number of flowers outlining the edge of the tree is **29**.
(2) Each of the 29 flowers on the outer edge has **9** petals.
(3) The branches from the trunk to the 29 flowers number 42 (21 to the left, 21 to the right), of these 22 are marked with a 'knot' at the midpoint (11 on the left, 11 on the right). The branches between bottom 22 flowers, and only counting ingoing (in a traversal up the tree), number 22, 11 on the left and 11 on the right. The top 7 flowers will count all branches as ingoing, thus 14 branch counts. The total branch count is **78**.

Note that 2x21+2x11+2x7=42+22+14=56+22+64+14=78, where the different groupings relevant to the set of alpha numbers is shown. Notice that we have the decomposition of 78=4x14+22 that agrees with the Tarot deck (with its 4 suits). We also have a decomposition 78=64+14 which reduces to 64 elements if the nature of the divination excludes the possibilities indicated by the 14 omitted elements, thus allowing for a constrained divination on 64 elements.

(4) There are **22** branches with knots at midpoint (as mentioned).

(5) 14 flowers have **4** incoming branches (4-chirality or suits). Of the other 15 flowers, the 8 on the sides have 1 incoming branch each, while the 7 on top have 2 incoming branches each (by chosen convention). Thus 4x14+8+2x7=56+8+14=64+14=78, and we again get the alpha numbers. Note, however, that in this branch counting we can differentiate according to "56+8+14=64+14" which means we must have a decomposition "**56**+8=**64**", i.e., the minor arcana (numbering 56) can be combined with 8 of the major arcana (the 8 gendered?) to arrive at a divination sub-system equivalent to the standard 'non-extended' I-Ching with its 64 hexagrams. Similarly, the decomposition "78=64+14" suggests an extension of the I-Ching formalism to the divination representation of the Tarot, but what would the 14 extensions be? It turns out 2 of the 14 'extensions' are in common usage, involving and extra line on certain symbols known as a 'yong' line. If this extension were done will all such symbols it would generate the 14 indicated. Also note that 64 comes from adding left and right, thus 64=**32**x2.

(6) The are **12** flowers with 'base knots'.

(7) Let's count the petals associated with the 14 'chiral' flowers (that have 4 incoming branches), there will be 9x14=126 such petals. Note the large central flower (at the top of the trunk) it is not only different in size, but has 11 petals, not 9. If the large central flower is included we get a total **137** from the 14 chiral flowers and the special central flower.

(8) Let's count the petals associated with the 15 non-chiral flowers: 9x15=135. W now come to resolution of an oddity in the diagram, the two circular elements at the base of the trunk of the tree seemingly connected to the bottom-most flowers on either side. If we now interpret that as extra petals from the left and right base flowers, then we count 2 addition petals, for a total of **137**.

The Assyrian Tree of Life is more explicit in detail, and more explicit in the numerology that is encoded. We have the following: {4,9,12,22,29,32,56,64,78,137}, which is the full set of 'alpha number' aside from alpha itself.

Was the tree of life silfium?

In ancient Mesopotamian mythology, the Tree of Life is often depicted as a sacred tree that symbolizes immortality and divine sustenance. It appears in various forms in Sumerian, Akkadian, and Assyrian art and literature. In the Bible, particularly in the Book of Genesis, the Tree of Life is mentioned in the Garden of Eden, representing eternal life and the divine presence of God.

Silphium (Ferula drudeana) was a plant that grew in the region of Cyrene (modern-day Libya) and was highly valued in the ancient world for its medicinal and culinary properties. It was particularly known for its use as a contraceptive and for various other health benefits. Silphium was so prized that it was harvested to extinction by the 1st century AD. Its image was even depicted on Cyrenian coins, symbolizing its importance to the economy and culture of the region. Some scholars and researchers have speculated that the Tree of Life could be symbolically linked to silphium due to its life-giving properties and its association with fertility and health. The idea of a tree that provides sustenance and life could metaphorically align with the benefits attributed to silphium. The use of plants and trees in ancient symbolism often carried multiple meanings, including life, death, fertility, and immortality. Thus, while there is no direct evidence to definitively link the Tree of Life with silphium, the symbolic connections are intriguing [271-273].

7.2.2 Hebrew Tree of Life Encoding

The Tree of Life in Kabbalistic tradition is a symbolic representation of the structure of the universe and the relationship between the divine and the material world. It consists of ten interconnected spheres (sephirot), each representing different attributes of God and stages of creation [274-281]. The sephirot are:

1. **Keter (Crown)** - Divine Will
2. **Chokhmah (Wisdom)** - Creative Insight
3. **Binah (Understanding)** - Analytical Thought
4. **Chesed (Kindness)** - Love and Mercy
5. **Gevurah (Severity)** - Judgment and Strength
6. **Tiferet (Beauty)** - Harmony and Compassion
7. **Netzach (Eternity)** - Victory and Endurance
8. **Hod (Glory)** - Humility and Surrender
9. **Yesod (Foundation)** - Connection and Communication
10. **Malchut (Kingdom)** - Manifestation and Reality

The Hebrew Tree of Life and Kabbalistic Number encoding

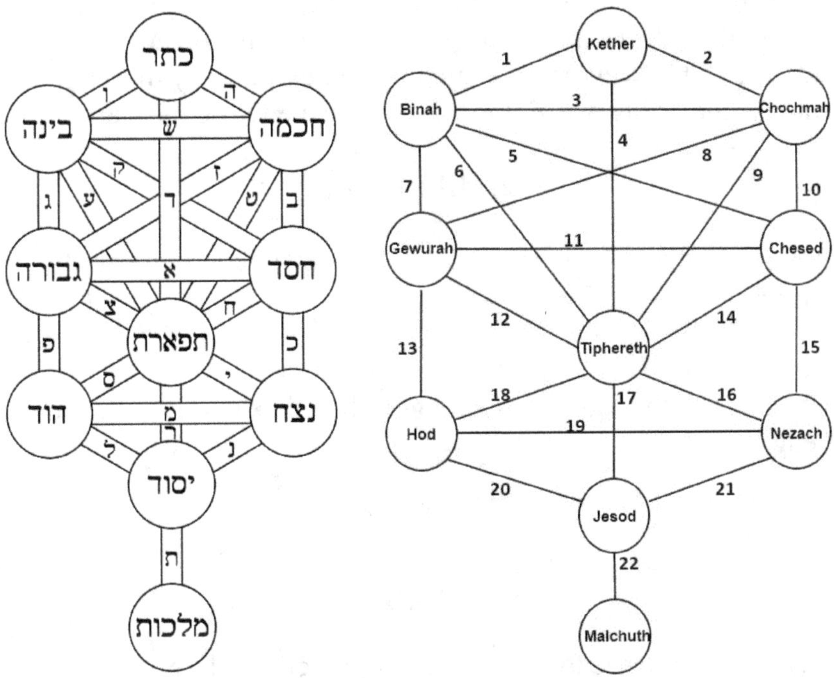

Fig. 7.3 The Kabbalah Tree of Life

Let's examine the diagram to see if the Kabbalistic numbers {10,22,78,137} are present. There are 10 nodes (sefirot, with names shown). There are 22 edges connecting the nodes (the 22 letters of the Hebrew alphabet. Let's now count the two-edge possibilities.

The Two-Edge Possibilities (78)

Let's consider the two-edge combinations that can be formed by 'descending'. Starting with edge labeled '1', we see that it connects (descending) to edges 3, 5, 6, and 7. We thus have the two-edge instances:

1-{3,5,6,7}

(e.g., two-edges 1-3, 1-5, 1-6, and 1-7). Likewise, starting at edge '2', we have:

2-{3,8,9,10}.

So far we have considered two-edges that are in a descending configuration (thus 1-2 not allowed, etc.). We now find the main exception to this counting rule and it has to do with two-edges that involve the central node (labeled 'tiphereth'). A two-edges entering tiphereth from 'above' can rise on the second edge. This means two-edges {4,6,9,12, and 14} enter tiphereth and can exit on any of the other 7 edges available (including ascending). Thus, for the two-edge starting with '4' we have:

4-{6,9,12,14,16,17,18}.

For the horizontal edge '3' we can descend on either side, there are thus six possibilities:

3-{5,6,7,8,9,10}.

For edges {5,8,7,10} that don't descend to tiphereth (directly), there are three two-edges each:

5-{11,14,15},
8-{11,12,13},
7-{11,12,13},
10-{11,14,15}.

Edges {6,9,12,14} do descend directly into tiphereth, so like '4', will have seven possible two edges each:

6-{4,9,12,14,16,17,18},
9-{4,6,12,14,16,17,18},
12-{4,6,9,14,16,17,18},

14-{4,6,9,12,16,17,18}.

For the horizontal edge '11' we can descend on either side, there are thus four possibilities:

11-{12,13,14,15}.

We now find the second exception to this counting rule, again for two-edges that involve the central node ('tiphereth'), but now we consider two-edges where the second edge ascends into tiphereth. This will be allowed if the overall two-edge describes an overall descent . Thus two-edges can entering tiphereth from 'below' on a rising second edge. This impacts two-edges that start with {13,15}. The exception cases will be listed in parentheses:

13-{(18),19,21},
15-{(16),19,20}.

The remaining edges are simply:

16-{19,20},
17-22,
18-{19,21}.

The total number of two-edges thereby indicated is 78.

The Three-Edge Possibilities (137)

We now need to modify the descent rule to 'descent in world level' (Net) and rise via tiphereth allowed, but only to sefirot below starting sefirot in three-edge. Let's descend via edge '1', there are then three edges that connect to sefirot other than tiphereth (for which there is the special rule). So, three-edges starting 1-7-X, 1-3-Y, 1-5-Z, then continue with X={11,12,13}, Y={8,9,10}, and Z={11,14,15} according to the normal descending rules.

Consider the three-edges "1-6-W". Edge '6' takes into Tiphereth, but it's not the first edge doing so. Do the special rules for Tiphereth only apply to three-edges with entry on the first edge, the minimal exception consistent with the two-edge exception? YES. Thus the allowed edges 'W' must be descending from Tiphereth (since no exception on last edge *leaving from* Tiphereth): W ={16,17,18}. There are, thus, 12 three-edges that start from edge '1'. Descending on the right starting with '2' we have a symmetric set of three-edges, thus also 12 in number.

Consider the three-edges starting with edge '4'. The exceptions for this allow all connection possibilities via Tiphereth (as with the two-edges). Consider the three-edge that starts 4-9-Z (e.g. the first two edges connect to the sefirot named Chochmah). The possibilities for Z are then {3,8,10}, where edges {8,10} descend from Chochmah, while edge '3' connects 'at the same level' horizontally over to the sefirot Binah. The starting node, Kether, is above the three final sefirot indicated by the final connections {3,8,10} so a rule of simple descent would allow all three possibilities. The rule, however, is required descent in 'World Order', where the diagram is split into four levels or 'Worlds': World IV consists of the top three sefiroth: Kether, Binah, Chochmah; World III consists of sefirot Gewurah and Chesed; World II consists of Tiphereth, Hod, and Nezach; and World I consists of Jesod and Malchuth. Thus, 4-9-3 is not allowed since it describes an overall descent from Kether to Binah, which is not a descent in World level (net). Thus, we have Z={8,10}. The three-edges starting from '4' are thus:

4-9-{8,10}
4-6-{5,7}
4-14-{11,15}
4-12-{11,13}
4-16-{19,20}
4-17-22
4-18-{19,21}

There are, thus, 13 three-edges that start with edge '4'.

Consider the three-edges stating with edge '3',we have:

3-5-{11,14,15}
3-6-{16,17,18}
3-7-{11,12,13}
3-8-{11,12,13}
3-9-{16,17,18}
3-10-{11,14,15}

There are, thus, 18 three-edges that start with edge '3'.

Let's now consider edges starting with 5 ,6, or 7 (the cases 8,9, 10, will be the same in number by symmetry):

5-11-{12,13}
5-14-{16,17,18}
5-15-{19,20}
6-14-{11,15}

6-12-{11,13}
6-16-{19,20}
6-17-22
6-18-{19,21}
7-12-{16,17,18}
7-13-{19,21}

There are 21 such three-edges, doubled when counting three-edges starting with 8,9, and 10.

Starting with the horizontal edge '11' we have on the left-hand side:

11-12-{16,17,18}
11-13-{19,21}

Also on the left-hand side, starting with edges 12 and 13, we have:

12-16-{19,20}
12-17-22
12-18-{19,21}
13-19-20
13-21-22

The total count for three-edges starting with 11 going to LHS, or with 12 or 13, is 13. If we count three-edges starting with 11 going to RHS or edge 14 or 15, we also get 13, by symmetry.

The three-edges starting with 16, 18, and 19 are simply enumerated:

16-19-21
16-20-22
18-19-20
18-21-22
19-20-22
19-21-22

The three-edges starting at 17, 20, 21, 22 can reach the bottom node via the first or second edge, so a boundary rule is needed. The rule is simply that we will count these abbreviated "three-edges" (with non-edge replaces with the '#' symbol):

17-22-#
21-22-#
20-22-#
22-#-#

With these rules the number of three-edges is: 37+22+42+26+6+4=137.

7.3 Divination methods indicated

The Kabbalistic numbers {10,22,78,137} are clearly present in the Hebrew Tree-of-Life. Is there a 56-step divination 'game' as has been outlined in the Game of Ur? There are too many possibilities to make a convincing argument on the basis of he diagram alone, however, in the Game of Ur we had 7 'markers' seemingly so as to perform 7 'loops' through the game-board. If we restrict to traversing the Hebrew Tree of Life in 7 loops, then an obvious 56 step result is found.

7.4 Chronology of ancient forms and practices of divination

Divination is the practice of seeking knowledge of the future or the unknown through supernatural means. It has been a part of human culture for as long as language, with various forms and practices emerging in different civilizations.

7.4.1 Interpretive Divination (no Alpha game, no process of refinement)

7.4.1.1 Augury

Augury is the practice of interpreting the behavior of birds to gain insight into future events. Ancient Romans and Greeks often used this method, observing the flight patterns, calls, and feeding habits of birds [282,283].

7.4.1.2 Geomancy and Bone (drop) Divination

Geomancy, or "divinatory geomancy," involves interpreting markings on the ground or patterns made by tossing soil or sand. This practice has roots in various cultures, including African and Arabic traditions [284,285]. In various ancient cultures, including those in Africa and Asia, divination was also performed using animal bones. The bones were cast or arranged in specific patterns, and the resulting configurations were interpreted to gain insights into the future or to seek guidance from ancestors [286,287].

7.4.1.3 Scrying

Scrying is the practice of gazing into a crystal ball, mirror, or other reflective surfaces to receive visions or insights. This method has been used in various cultures, including ancient Egypt and medieval Europe [288,289].

7.4.1.4 Oracle Bones

Oracle bones were used in ancient China during the Shang Dynasty (c. 1600–1046 BCE) for divination. Questions were inscribed on bones or turtle shells, which were then heated until cracks formed. The patterns of the cracks were interpreted as answers from the ancestors or deities [290,291].

7.4.1.5 Dream Interpretation

Dream interpretation has been practiced since ancient times, with cultures like the Egyptians and Greeks believing that dreams could provide prophetic insights or messages from the divine. More recently, its become an area of psychological exploration [292].

The divination Methods in the sections 7.4.2 -- 7.4.7 that follow have the potential for refinement and can therefore be refined to be an 'Alpha' game, where the emergent Alpha numbers are exhibited. These more sophisticated divination methods will be seen to propagate from origination (Chinese Siberia/Siberian Steppe or Babylon, for example) in a 'wave' that progressively reaches different cultures/civilizations further away.

7.4.2 First Wave: Yarrow Sticks

The first wave of divination practices involved the use of yarrow sticks, an ancient method originating in China. This method is closely associated with the I Ching, or Book of Changes, which is one of the oldest Chinese classic texts. The use of yarrow sticks emphasizes the connection between nature and divination, as the sticks are derived from a natural source [293] In this divination process, the structural numbers include many of the familiar 'Alpha numbers'.

Yarrow sticks divination, also known as **"Yijing"** or **"I Ching"** divination, is one of the oldest forms of divination in Chinese culture, dating back to at least the Shang Dynasty (c. 1600–1046 BCE). The practice involves the use of 50 yarrow sticks, which are manipulated in a specific way to produce a hexagram—a figure composed of six lines, each of which can be either broken (yin) or unbroken (yang).

Process:

1. **Preparation**: The practitioner begins with 50 yarrow sticks. One stick is set aside, leaving 49.
2. **Divination**: The remaining sticks are divided into two piles, and the number of sticks in each pile is counted. This process is repeated several times, with the number of sticks in each pile determining whether the resulting line of the hexagram is broken or unbroken.

3. **Hexagram Formation**: After six rounds of counting, a hexagram is formed, which is then interpreted using the I Ching text. Each hexagram has associated meanings and can provide guidance on various life situations.

The I Ching is not only a divination tool but also a philosophical text that reflects the principles of change and balance in the universe. It emphasizes the dynamic interplay of yin and yang and the importance of adapting to changing circumstances.

While yarrow sticks divination is one of the oldest documented forms, there are other ancient practices that may predate it, although they are less well-documented:

7.4.3 Second Wave: Livers

The second wave of divination involved the examination of animal livers, a practice known as hepatoscopy. This method was prevalent in various ancient cultures simply as scrying initially, but under the Etruscans and Babylonians this became much more elaborate. In this practice the location of a blemish on the liver would be noted with reference to a liver-reference ceramic figurine. Some of these figurines show a clear grid pattern to identify where the blemish resided, and there are 56 grid cells in such 'student trainer' examples, thus the Alpha numbers are already evident in these early divination methods.

7.4.4 Third Wave: I-Ching Hexagrams

The third wave of divination is characterized by the explicit use of I-Ching hexagrams, which are formed by casting yarrow sticks or coins. Each hexagram consists of six lines, which can be either broken or unbroken, representing yin and yang. The interpretation of these hexagrams provides a framework for understanding the dynamics of change and the interplay of opposites in life [294]

7.4.5 Fourth Wave: Use of Dice

The fourth wave introduced the use of dice as a tool for divination, known as astragalomancy. This practice involved casting dice to obtain random numbers, which were then interpreted according to established systems. Dice divination was common in various cultures, including ancient Rome and Greece, and was often associated with chance and fate.

7.4.5.1 Runes

Runes are characters from ancient alphabets used by Germanic tribes. They were often carved on stones or wood and used for divination by casting or drawing them and interpreting their meanings [295,296].

7.4.5.2 The Game of Ur

In the case of Tarot we have a divination process [297] using 56 minor arcana and 22 major arcana. Kabbalah is also thought to have had a divination process based on its tree of life diagram, but there is no consensus on how this divination was done. Certain versions, however, traverse the tree repeatedly such that 56 steps are made. Preceding both Tarot and kabbalah, and passed to us now only via discovery of the Royal Game of UR gameboard is the following:

Fig. 7.4 The Royal Game of Ur Gameboard [298].

It is not known how the Royal Game of UR was played [20], but some versions, notably, involve 56 steps.

In addition to the well-preserved game-board shown above, each player (2) had 7 markers or tokens (different colored sets). The path of the game-board of traversal is not known nor the manner of that traversal. There are hints from versions of the game-board found in Egypt (this is not the only game transferred to ancient Egypt by the even more ancient Sumer) in the form of sketches of the board made into stone by Egyptians at pyramid work sites.

In the sketch below is the traversal I believe was used, consistent with the kabbalistic numbers outlined previously. In terms of adjoining sequential squares traversed like in a modern-day game-board, there are 31 steps, 32 if you count the last step off the game board to the 'imaginary' 4th node indicated upon exit. Other ideas on traversal have been proposed with paths having 14 steps (Bell-Finkel); 16 steps (Masters); 23 steps (Skirynk); 27 steps (Murray); in addition to the '31 steps' outlined below.

This counting of 'steps' is misleading, however, since they imply motion like in a modern day board-game where if you roll a 3 (on a 4-sided die), then you move three spaces. What is proposed instead is that you 'move' by selecting the chirality in the next 4-suited selection. In other words, if you roll a 3 then you move to the next '3'. If traversal is made by selection on chirality eight times, then a complete board traversal is made. In other words, if the {1,2,3,4} cycles are counted on the traversal path indicated, then there are eight such cycles. There are thus eight chiral 'moves' made by a marker as it traverses the board, and to traverse all seven markers that a given player has, this then entails 56 chiral (4-suited) steps. We arrive at the critical kabbalistic number on number of moves of 56, each with the 4-suited, or doubly chiral, characteristic. This divination numerology will reappear with the Egyptian Malmut Tarot deck (56 cards, 4-suited) that is thought to have arrived from Sumer via the contemporaneous Babylonians of that time.

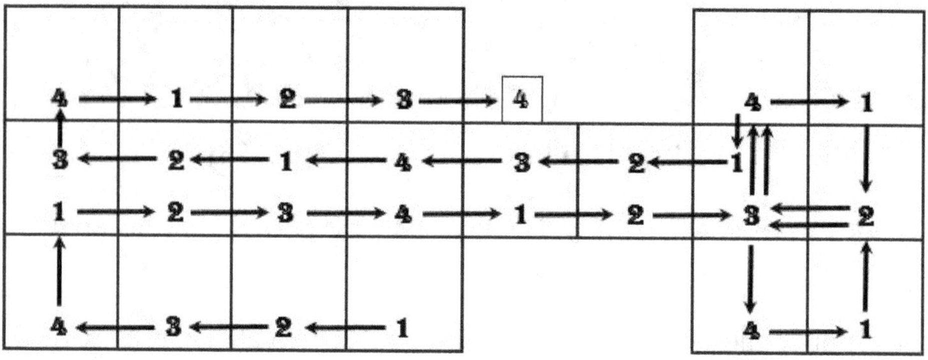

Fig. 7.4 The Royal Game of Ur Gameboard traversal in 56 steps.

7.4.5.3 Possible Divination Process with Hebrew Tree of Life Diagram

We will make seven loops, each starting and ending at Kether (the top node). Each loop, starting at Kether, will loop around Tiphereth (counterclockwise), with each loop omitting a different sefirot on the loop by passing thru Tiphereth instead. The edges traversed in each loop are thus as follows:

Loop 1: {4,12,13,21,20,15,10,2} (skips Binah)
Loop 2: {1, 6, 18,21,20,15,10,2} (skips Gewurah)
Loop 3: {1, 7, 12,17,20,15,10,2} (skips Hod)
Loop 4: {1, 7, 13,18,16,15,10,2} (skips Jesod)

Loop 5: {1, 7, 13,21,17,14,10,2} (skips Nezach)
Loop 6: {1, 7, 13,21,20,16, 9, 2} (skips Chesed)
Loop 7: {1, 7, 13,21,17,15,14,4} (skips Chochmah)

7.4.6 Fifth Wave: Use of Sheet Dice (Tarot)

The fifth wave marks the emergence of Tarot cards, which can be seen as a form of sheet dice. Tarot decks consist of 78 cards, each with symbolic imagery and meanings.

Origins of Tarot

Misconception #1: Correlation of Fall of Byzantium, with immediate rise of Tarot in West, indicates causation: Byzantium was 'blocking' Tarot in favor of its established Zodiac-based divination method.
Correction #1: The timing of the rise of Tarot had more to do with the invention of paper, and transmission of cartomancy during the Mameluk Sultanate, which happened (in the West) to mainly 'transmit' at the time of the Fall of Byzantium.

Misconception #2: That Tarot was derived from Kabbalah, thus it's early appearance in Italy ~1450 should associate with the Jewish community, some of which practiced (Jewish) Kabbalah, which, in turn, derives from Babylonian Kabbalah (that was acquired by the Hebrews during the Babylonian captivity where they were valued slaves as scribes).
Correction #2: The early appearance of Tarot was not in ~1450 Italy, as commonly believed, but centuries earlier in Spain (via Moors) and anywhere in in ~1350 Europe that had trade with Egypt (under the Mamaluk Sultante). This historical correction makes it very unlikely the Kabbalah numerics in Tarot derived from Jewish Kabbalah as assumed, remarkably, it appears to have derived directly from Babylonian Kabbalah.

Misconception #3: Origin of Tarot was Egypt under Mamaluk Sultanate, transmitted via four paths into Europe: (i) Gypsies in Balkans; (ii) Moors in Spain; (iii) Crusaders returning the Western Europe; and (iv) Central and Northern European trade with Egypt directly (and buying the 'card game'). And, further, that the origin in Egypt runs to ancient Egypt (presumption of Crowley and his predecessors to be discussed below).
Correction #3: Although transmitted via Egypt under the Mamaluk Sultanate, the roots of Tarot were not Egypt, but Babylonian (Persian).

7.4.7 Sixth Wave (from Babylon): Use of Zodiac

The sixth wave of divination practices involved the use of the Zodiac, which originated in Babylonian astrology. The Zodiac consists of twelve signs, each associated with specific traits and influences. Astrologers would interpret the positions of celestial bodies in relation to the Zodiac signs to provide insights into human behavior and future events [299]

7.4.8 Seventh Wave (from Ottoman/Istanbul): Zodiac

The seventh wave continued the tradition of Zodiac-based divination, particularly within the context of Ottoman astrology. Ottoman scholars expanded upon earlier astrological practices, integrating them with Islamic teachings. This wave emphasized the importance of celestial movements and their influence on earthly affairs [300]

Western astrology is based on Ptolemy's *Tetrabiblios* (2nd century), which has its roots in Hellenistic and before that Babylonian divination/astronomy traditions. Many of the constructs in the astrology zodiac are of distinctive Babylonian numerological origin (consisting of the number 60 and its factors and multiples). Thus, the zodiac has 12 divisions of the sky, each of 30 degrees, comprising a 360 degree sky. The twelve signs are aligned with the four elements. In one form or another, this astrological (horoscope-based) method for divination traces as far back as 19th century BCE Mesopotamia, with spread from there to Greece, Rome, the Islamic World, and then Europe [301]. During the Enlightenment astrology would lose its status as an area of legitimate scholarly pursuit [302].

7.4.9 Eighth Wave: Emergence of Tarot in the West

The eighth wave highlights the emergence of Tarot in Western culture, particularly during the Renaissance. Tarot cards began to be associated with mysticism and esoteric traditions, leading to their use in divination. This period saw the development of various Tarot systems, each with unique interpretations and symbolism [303]. The emergence of Tarot in the West was marked by the transition from a card game to a tool for divination and self-reflection. This transformation was influenced by various mystical and philosophical movements, including Hermeticism and Kabbalah. The Tarot became a means of exploring the subconscious and understanding the human experience through archetypal imagery [304,305].

One of the descendants of Kabbalah in the West (in Medieval times) is Tarot. Thus, it should come as no surprise that the Tarot deck should have 56 cards (of one type, the minor arcana), and another 22 of another type (the major arcana), their total being the recurring Kabbalistic number 78 [306]. Note: the familiar card deck is based on the 56-4=52 cards, where the gender-pairing with the 4 Jacks (the 4 Maidens) have been dropped. (There is a tarot-deck equivalent game known as "Old Maid" where the 4 cards are put into the deck, the old maids.)

In the case of Tarot we have a divination process [297] using 56 minor arcana and 22 major arcana. Kabbalah is also thought to have had a divination process based on its tree of life diagram, but there is no consensus on how this divination was done. Certain versions, however, traverse the tree repeatedly such that 56 steps are made. Preceding both Tarot and kabbalah, and passed to us now only via discovery of the Royal Game of UR gameboard is the following:

7.4.10 Ninth Wave: Tarot and Kabbalah

The ninth wave explores the relationship between Tarot and Kabbalah, a mystical Jewish tradition. In the late 19th and early 20th centuries, occultists began to draw parallels between the Tarot's structure and Kabbalistic teachings. The Major Arcana cards were often associated with the Sefirot (divine emanations) of the Kabbalistic Tree of Life, creating a framework for spiritual exploration and understanding [307,308].

Note: the familiar card deck is based on the 56-4=52 cards, where the gender-pairing with the 4 Jacks (the 4 Maidens) have been dropped. (There is a tarot-deck equivalent game known as "Old Maid" where the 4 cards are put into the deck, the old maids.)

When kabbalah was translated to Western esoteric systems the result is what is known as Qabalah (or Hermetic Qabalah). Qabalah provides the underlying framework for magical societies such as the Golden Dawn and the Thelemic Orders [309]. The most recent incarnation of Qabalah is what is known as 'English Qabalah'. According to Stratton-Kent [310,311] with comment indicated in square parenthesis, "the English Qaballa is a qabalah and not a system of numerology. A qabalah is specifically related to three factors: one, a language; two, a 'holy' text or texts; three, mathematical laws [gematria-like rules] at work in these two."

Chapter 8. The Mystery of the Logos

In this chapter the mystery of the Logos is explored, but in the context of the what has been revealed about the mystery of Alpha, from Emanator Theory, in prior chapters. It will be shown in Sec. 8.4 that the Logos describes a "word-centric" reality or a "narrative" reality, with a variety of reifications and incarnations possible, and this is precisely what would be expected from the MIE maximization of complexity with continuity from prior chapters. Just like early identification of 1/Alpha=137 and other Kabbalistic numbers (that are also 'alpha numbers') is seen in (divination) games-of-refinement, there is also early recognition of the nebulous concept of Logos. So what is the Logos? The classical notions and descriptions of the Logos are given in Section 8.1. In Section 8.2 the analysis "goes meta". The starting point is Coleridge, who came to the topic of the Logos late in his writing career. On numerous occasions Coleridge indicates that all of his writings were of the Logos in one form or another (to his own surprise, upon discovery and his own understanding of the Logos). Section 8.2 does text analytics analysis of the entire works of Coleridge (and Shakespeare and Machiavelli). We find that the Logos derives from individual knowledge in a variety of ways, but also from cultural knowledge, starting with early symbolic manipulation (proto-civilization) and written language manipulation (civilization). Given this context, Section 8.3 describes the Logos' cultural developments from proto-civilization through to modern civilization, with attention to the written retention of knowledge (mainly concerning law and religion, where early science was considered religion).

Starting with symbolic manipulation (including cave art and early flutes) for the proto-civilizations, we trace backwards to the beginnings of the Logos in the (Siberian) Steppe (consistent with the Out-of-Steppe Hypothesis indicated in Chapter 6). The waves of settlement from the Steppe went from the Siberian East westward into northern Europe, and with branches southward taken wherever mountain passes permitted. Thus, directly south from Siberia was done, through China and into southeast Asia (from that side of the Himalayas). Likewise, if the Steppe is followed to central Asia, then a southward is possible via the Khyber Pass into the Indian subcontinent. Following the Steppe further westward (over a few thousand years of migration, for example) we arrive at the Pontic-Caspian steppe -- especially the region bordering the Black Sea from the north. This location would eventually be the starting point for a hypothetical Proto-Indo-European (PIE) civilization. Most of the literature associates this hypothetical proto-civilization with the Kurgans (named after burial mounds known as kurgans, from 4500 BC). However, given the evidence of Gobekli Tepe (described in Section 8.3.2), which dates to 9600 BC, we see that we have proto-civilization in the Black Sea area, on the southside as early as 9600BC, and if originating from the north, then something on the north of the Black Sea from 10000 BC or earlier. This is when the Black Sea would have been Blue, however, so Section 8.3.1 moves the critical time point for first PIE proto-civilization to 10000BC on the north shore of the "Blue Sea" (so this is the "Blue Sea Hypothesis").

Meltwater Pulse 1B occurred 9500BC, but did not breach the Bosphorus and change the Blue Sea to the current Black Sea until around 7500BC, so the existence of a proto-civilization on the north shore of a massive freshwater lake (the Blue Sea) seems likely. From this, the ancient civilizations and proto-civilization of Eurasia describe the development of the Logos along numerous parallel threads that form the basis of modern society (Section 8.3.3). Section 8.4 indicates a description of the Logos as encompassing a description of 'Narrative Reality'. If taken from the perspective of an individual, how is 'narrative reality' different from consciousness? Maybe it isn't (and this is the topic of the next chapter). Evidently the mystery of the Logos is partly shrouded in the mystery of the nature of consciousness itself, on an individual level, and the nature of civilization on a cultural level.

8.1 The Classical Logos (Hellenic, Jewish, Christian, Islamic)

The concept of the "Logos" has a rich and varied history, spanning philosophy, theology, and literature. It originates from ancient Greek philosophy (with some roots even earlier than that) and has been interpreted in various ways by different thinkers throughout history. In ancient Greek philosophy, particularly in the works of Heraclitus, the Logos refers to a principle of order and knowledge. Heraclitus used it to describe the rational structure of the universe, suggesting that there is a reason or logic that governs all things.

The Stoics further developed the concept, viewing the Logos as the divine reason or rationality that permeates the cosmos. They believed that the

Logos was a guiding force in nature and human affairs. In Christian theology, particularly in the Gospel of John, the Logos is identified with Christ. The famous opening lines of John 1:1 state, "In the beginning was the Word (Logos), and the Word was with God, and the Word was God." Here, the Logos represents the divine reason and creative order through which the world was made. In Sufism, the Logos can be understood as an inner connection to God. Sufi thinkers, such as Ibn Arabi, have often equated the Logos with the idea of the "Universal Man" (al-Insan al-Kamil), who represents the archetype of God's presence in the world and the perfect realization of divine attributes in a human form [312].

The "mystery of the Logos" includes the esoteric understanding of this concept, particularly in relation to its implications for existence, creation, and the divine. It encompasses the idea that the Logos is not just a rational principle but also a profound mystery that connects humanity to the divine.

Descriptions of the Logos by Philo, St. John, and Coleridge:
Philo of Alexandria: A Hellenistic Jewish philosopher, Philo integrated Greek philosophy with Jewish theology. He described the Logos as an intermediary between God and the world, a divine reason that helps to explain the creation and order of the universe.
St. John the Evangelist: In the New Testament, John presents the Logos as both a creative force and a personal being in the form of Jesus Christ.
Samuel Taylor Coleridge: The Romantic poet and philosopher Coleridge engaged with the concept of the Logos in his writings. He viewed the Logos as a unifying principle that connects the material and spiritual worlds. In his work, he often explored the relationship between imagination, reason, and the divine, suggesting that the Logos embodies the creative and rational aspects of existence. Coleridge believed that the Logos is essential for understanding the unity of all things and the divine purpose behind creation. A text analysis of the entire works of Coleridge will be described in what follows, where the attributes of 'The Logos" will be extracted.

8.2 Text Analytics identification of the Logos [11]

8.2.1 Word frequency Analysis: Machiavelli's Polysemy on *Fortuna* and *Virtu*

Text analysis is done for Il Principe (The Prince) by Machiavelli. The analysis of Il Principe is done on the original Italian text and reveals clear use of polysemy by Machiavelli for the concepts *fortuna* and *virtu*.

Furthermore, the polysemy usage for *virtu* is clearly linked to *fortuna* (e.g., it is *virtu* to recognize *fortuna* and be prepared to seize it). The importance of polysemy and the errors in translation of texts that use it, such as translations of Machiavelli's Il Principe (from ancient medieval Italian) have been noted in variety of translation contexts, including recently for Spanish [313,314]: "During the Middle Ages the term virtu experienced an important change in meaning, from the initial idea of 'maturity' or 'excellence', to that of 'perfection' in a moral sense. Machiavelli made the idea of virtu a key concept, but bestowing its initial polysemy on it and, therefore, disregarding its moral dimension. Translations into Spanish that did not take into account that polysemy would betray a fundamental idea in works as relevant as The Prince."

The counts (shown in Table 5.1 of [11] with further detail) reveal four large classes of words other than generic language constructs (e.g., conjunction, pronouns, prepositions, adjectives, adverbs):

Type I: power, and large entities of power, such as The Prince, The People, The State, and power: Principe, popolo/uomini, Stato/stato, potere.
Type II: concerns opportunity, preparedness, and timing: fortuna, virtu, tempo, tempi, volta.
Type III: players/agents/groups such as subjects, soldiers, powerful, city, (small) principality, army, enemies, friends: sudditi, soldati, potenti, citta, principato, esercito, nemici, amici.
Type IV: actions such as 'actions' (azioni), war (guerra), force (forza), cruelty (crudelta).

Discussion surrounding type I and type III words is to delineate and understand the groups and entities present. This is a structural part of the political science discourse and makes frequent reference to the role of the (new) Prince, and in advising the new Prince (the stated objective of the text). Discussion surrounding type II words focuses on how to be successful through preparedness, awareness, initiative, and use of timing. Discussion surrounding type IV and some type III words were concerned with how to manage risk (regardless of appearances in some circumstances).

To explore further, let's get a visual on the high frequency 'important words', where high-count anomalous word parings will be visually apparent. We will simply map each word of the text to be simply"-"

unless it's one of the high frequency words listed above. The first part of Il Principe book then compresses to:

____stato__tempo__________________
_______________________________________stato____
____fortuna__
_______________________________uomini____
____
____
____tempo__
____
____
____
____
____
___________________________uomini____
____
____
_____________________uomini____
____
____P
rincipe____Principe____
____
____stato__________________________________
____tempo____________________________________
__stato______________________________stato____
__________tempo____
_________________________Stato____
___________________stato____
____Principe______________________________
_Principe____
___________________uomini____
____tempo______________________
____
____
____
__________fortuna____fortuna__________
____potere_uomini_____________
_____________Principe____tempo_potere_______________________________Stato__P
rincipe_________________________________Principe___________________________fortuna_
_virtù____________

For the above output we end with the first of the fortuna-virtu word pairings, there are 5 with that order, and 6 with the reverse order. Translations of these passages using Google Translate, with the untranslated fortuna and virtu terms (to retain polysemy), results in an interesting distillation of one of the main messages of Machiavelli (see [11] for further details).

8.2.2 Word frequency analysis: Coleridge's Hidden Polysemy on *Logos* [11]

Word frequency analysis can catch polysemy, as with Machiavelli's Il Principe, but can also identify high-frequency word subsets indicative of the topic discussed. In this section an attempt to define the concept of the Logos using such an analysis by analyzing the entire works of Coleridge will be done for this purpose as will be explained in what follows.

Early Stoicism describes the Logos as the active and vivifying power. Philo, A Hellenistic Jew (c. 30 BC) thought of the Logos as the aspect of the divine that creates the world and is "the bond of everything". The most famous description/definition of the Logos is described in two lines in the Book of John 1:1 and 1:14 (where Jesus is the incarnation of the Logos). Elsewhere in the Book of John, however, all other translations of logos are for unrelated common words, such as "saying". Thus, there are roughly four sentences of information to describe the core concept of "Logos". Centuries later, Coleridge (1772-1834) was not satisfied with this and sought a more clear exposition of the Logos.

To Coleridge the Logos presented a unifying philosophy/theology [315]. In Coleridge's earliest known mention of efforts to examine the Logos [316] he describes the Logos as "the communicative intelligence in nature and man". With the latter he seems to connect with earlier Kabbalistic notions of flow of information in nature and man. Coleridge also spoke of Logos "Human and Divine", where he argued for Faith as a form of human Logos: "Faith must be a Light originating in the Logos" [317]. For the last twenty years of his life Coleridge put time on his "Opus Maximum" on the Logos, but it was never completed, only being published as a collection of writings in 2002 [317]. Aside from his philosophical refinements to the Logos just mentioned, Coleridge was not able to advance his plan for an "Opus Maximum", on the Logos as a rational basis for Christianity, much beyond stating his objectives and restating previous ideas.

Coleridge describes his discovery of the concept of the Logos and his efforts to study it as occurring only after being a prolific writer and poet for decades [317]. Coleridge then makes an interesting point about his prior career and body of work in relation to his current efforts to explore the Logos – he says that all of his prior work was about the Logos in some manner without ever knowing of the explicit unifying concept. In essence, he is claiming that there is a "hidden polysemy" in his work where the Logos concept is scattered over many words, themselves repeated in various ways..... *So let's do a word frequency analysis on the entire body of work by Coleridge and see if we can distill the concept of "Logos" down to a collection of high frequency words with strong context associations that are "the Logos".*

High Frequency words extracted (see [11] for details), and subjunctive-type and romantic-type words are dropped, where subjunctive-type = subjunctive and subjunctive-related phrase linkage words; romantic-type = romantic style word usage, especially vagueness with subjunctive. Six categories result: heart, power, word, light, three, human:

Categories:

- heart+ = {heart, sense, feelings, feeling} (Sense)
- power+ = {power, whole, nature, world, natural} (Nature)
- word+ = {words, thought, think, truth, reason, written, language, thoughts, poetry, knowledge, genius, understanding, number} (Word)
- light+ = {light, spirit} (emanation)
- three+ = {three, faith, character, believe, Church, Christ, moral, Christian, religion, purpose, father, Spirit} (trinity)
- human+ = {human, death} (emanation incarnate)

The indications about Coleridge's earlier work being in-line with a study of the aspects of the Logos is correct. The categories "word+" and "power+" contain all the standard Greek translations of "logos" relevant to the concept. The categories "light+" and "human+" describe emanation phenomena in line with ideas of logos as a divine emanation. The category "three+" encompasses Coleridges efforts to affirm Trinitarian Christianity on a rational basis (later an explicit effort in his Opus Maximus [317], where he uses the Logos concept to provide a foundation for that rational basis). The one category that doesn't fit is the first encountered (highest count), where Coleridge speaks of the subjective senses via the terms {heart, sense, feelings, feeling}. The "heart+"

category may be merged with "human+", thereby describing both the subjective and objective perspectives of the emanation incarnate. This is consistent with his later writings on this in [317]. It would appear Coleridge's comments on his early work being entirely focused on the Logos, without his knowing it, is consistent with the word frequency analysis just shown. Furthermore, Coleridge's early works indicated the main outline of Coleridge's contributions to the Logos concept that would follow decades later.

The key terms describing the nebulous concept of the Logos, other than those pertaining to the individual (covered later), are those pertaining to law, religion, government administrations,… "civilization" in other words. Thus, we explore the development of civilization in Section 8.3 to track the intertwined development of the Logos in this social sense. Before moving on from the text analytics let's do one more analysis.

8.2.3 Shakespearean Phrases [11]

Let's consider the frequency on (consecutive) three-word groupings across the entire works of Shakespeare (Taken from the Project Gutenberg Library).

The top five most used three-word phrases:

1. 192 count on Iwillnot
2. 188 count on Iprayyou
3. 138 count on Idonot
4. 137 count on Iama
5. 132 count on Iamnot

Notice how the three word phrases are also all three syllable phrases. In analysis on specific plays, not based on all of Shakespeare's works (including the Sonnets, etc.), the most common three-word phrase is "I pray you". One of the most common two-syllable phrases (analysis not shown) is "Prithee". Shakespeare has many lines in plays that begin with the three-syllable, three-word phrase "I pray Thee", or with the similar meaning two-syllable "Prithee". This stylistic convention is one of Shakespeare's methods for "keeping the meter", i.e., for maintaining iambic pentameter.

Iambic pentameter is where you have five two-syllable steps per line of dialogue or verse, i.e., ten syllables per line. In practice eleven syllable lines were common. Furthermore, just as important as the meter was the accent on the syllables in that meter. Here distinctive styles emerge even more clearly between authors. Accenting rules on top of the iambic pentameter were strictly obeyed by some authors, like Pope and Shakespeare, and by others like Donne not at all (Ben Johnson is quoted as having said that Donne deserved hanging for this [318], so the artistic sensibilities regarding this matter could be quite heated). For the analysis to follow we will content ourselves with the task of identifying the iambic pentameter structure in the works of Shakespeare without further discussion of accent. Even so, this is not a simple matter as there is not a straightforward syllable look-up table for words.

Iambic phrase analysis

In [11] text analysis is done for Henry VI by William Shakespeare (and Christopher Marlowe). The analysis of Henry VI involves iambic profiling for Parts 1 and 2, revealing a shift in iambic profile, possibly associated with a shift in authorship. The application of the script and production of histograms showing syllables counts on lines is left to the exercises. Other than a main peak at 10 syllables, there are strong side-peaks at 9 and 11 syllables, and the difference in the histograms based on Henry VI Part 1 and Part 2 indicate a shift in authorship as expected (since this is known to be true in this case as Marlowe officially covered for Shakespeare at this time due to illness).

Note that a common syllable count was 11 then 10. In essence, the half step at the end of the first line creates tension, that is resolved with the second line giving an exact 10-syllable count. If the pause between lines is counted as its own ‘state’, then the common two line phrase consists of 22 states.

8.3 The Logos: From proto-civilization to Modern Era

8.3.1 The Blue Sea Proto-Civilization Hypothesis

Known Black Sea civilization artefacts and technologies

The Black Sea region has been home to various ancient cultures and civilizations, each contributing to a rich archaeological record. While there isn't a single "lost civilization" specifically identified as a Black Sea civilization, several ancient peoples thrived in this area, particularly during the Bronze Age and earlier.

Pottery is one of the most common archaeological finds and can provide insights into the daily life, trade, and artistic expression of ancient cultures. Different styles of pottery can indicate cultural affiliations and trade networks. Examples:

Thracian Pottery: Known for its intricate designs and use of local clay.

Greek Colonies: Pottery from Greek settlements around the Black Sea, such as those in modern-day Crimea and the Caucasus, often features black-figure and red-figure techniques [319].

The development of metallurgy, particularly in bronze and iron, was crucial for tool-making, weaponry, and artistic expression. Artifacts such as tools, weapons, and jewelry provide evidence of advanced metalworking techniques. Examples:

Bronze Age Tools: Swords, axes, and other tools made from bronze have been found in burial mounds (kurgans) associated with the Scythians and other nomadic tribes.

Gold Artifacts: The Thracians and Scythians produced exquisite gold jewelry and ceremonial items, showcasing advanced techniques in goldsmithing [320].

Properties of a hypothetical 'lost' Black Sea Civilization (if there was one)

Consider a black sea civilization lost at the time of its flooding, what technology and language type would it have likely had? The concept of a "lost civilization" in the Black Sea region, particularly in relation to the flooding associated with Meltwater Pulse 1B (approximately 14,600 years ago), is speculative. However, we can draw on archaeological and linguistic evidence from the surrounding cultures to hypothesize about the technologies and languages that might have existed in such a civilization.

Hypothetical Technologies

Maritime Technology: Given the Black Sea's significance as a waterway, a civilization in this region would likely have developed advanced boat-building techniques. This could include dugout canoes or plank-built vessels, which would facilitate fishing, trade, and exploration [321].
Agricultural Practices: The transition from hunter-gatherer societies to settled agricultural communities would have been crucial. Technologies such as the plow, irrigation systems, and storage facilities for grains would likely have been developed [322].

Pottery and Craftsmanship: Pottery would have been essential for storage, cooking, and trade. Techniques for firing clay and decorative styles would reflect the culture's artistic expression ([323] discusses the evolution of pottery techniques).

Metallurgy: If this civilization existed during the late Neolithic or early Bronze Age, it might have developed metallurgy, particularly in copper and bronze, for tools and weapons [324].

Hypothetical Language Type

Proto-Indo-European (PIE): If a civilization existed in the Black Sea region during the time of the flooding, it might have spoken a language related to Proto-Indo-European, the hypothesized ancestor of many modern languages in Europe and parts of Asia. The Black Sea is often considered a potential homeland for the speakers of PIE, particularly in the context of the Kurgan hypothesis [325].

Pre-Greek Languages: In the context of later periods, languages spoken in the region before the arrival of Greek settlers could include various pre-Greek languages, which are not well-documented but are believed to have existed in the area [326].

8.3.2 The proto-civilization Göbekli Tepe

Göbekli Tepe, located in southeastern Turkey, is one of the most significant archaeological sites related to early human civilization. Dating back to around 9600 to 9500 BCE, it predates the advent of agriculture and is often considered a proto-civilization due to its complex social and religious structures.

Göbekli Tepe features massive stone pillars arranged in circular and oval-shaped enclosures. The pillars, some weighing up to 20 tons, are intricately carved with reliefs of animals such as snakes, foxes, and birds. The site consists of at least 20 circular structures, with the most famous being Enclosure D, which contains T-shaped pillars [327]. The site is believed to have served as a ritual center, indicating that early humans engaged in complex social and religious practices long before the establishment of settled agricultural communities. The construction of Göbekli Tepe suggests a high level of organization and cooperation among hunter-gatherer groups [328]. Göbekli Tepe challenges traditional narratives about the development of civilization, suggesting that complex religious practices may have preceded and facilitated the transition to agriculture. This indicates that social and religious motivations could have played a crucial role in the development of settled societies [329].

Other Developments Around the Same Time

Çatalhöyük: Located in modern-day Turkey, Çatalhöyük is one of the earliest known urban settlements, dating from approximately 7500 to 5700 BCE. The site features densely packed mud-brick houses and evidence of early agricultural practices, including the cultivation of wheat and barley. The community is notable for its elaborate wall paintings and figurines, indicating a rich cultural life [330].

***Jericho**:* One of the oldest continuously inhabited cities in the world, Jericho, located in the West Bank, has evidence of settlement dating back to around 9000 BCE. The site features a large stone wall and a tower, suggesting early forms of urban defense and social organization. Jericho is also associated with early agricultural practices, including the domestication of plants and animals [331].

***Natufian Culture**:* The Natufian culture (c. 12,500 to 9,500 BCE) in the Levant represents a transitional phase between hunter-gatherer societies and settled agricultural communities. The Natufians are known for their semi-sedentary lifestyle, the construction of stone structures, and the early domestication of plants, particularly wild cereals [332]

Göbekli Tepe and other contemporary sites like Çatalhöyük and Jericho illustrate the complexity of social, religious, and economic structures that emerged long before the advent of agriculture, challenging traditional views of civilization's origins [333,334].

8.3.3 Civilization from Neolithic to Modern Era

Neolithic Revolution (10,000 - 4,000 years ago)

Transition from nomadic hunter-gatherer societies to settled agricultural communities. Domestication of plants and animals, development of pottery, weaving, and the establishment of permanent settlements [335]

Bronze Age to Iron Age (4,000 - 500 BCE)

Development of metallurgy, writing systems (details below), and complex societies. Emergence of city-states, trade networks, and early forms of government and religion [336].

Written Languages

The invention of writing marked a significant advancement in human communication, allowing for the preservation and transmission of knowledge across generations. Written language evolved from pictographs to complex scripts, enabling the documentation of history, law, and literature. Despite the diversity of writing systems, many share common features, such as the use of symbols to represent sounds or ideas. The development of alphabets, syllabaries, and logographic systems reflects a universal human need to record and communicate information efficiently [337].

The invention of writing systems was around 3200 BCE in Mesopotamia (cuneiform) and Egypt (hieroglyphics) and marked a significant advancement in communication, allowing for the recording of information and the transmission of knowledge across generations [338]. The development of alphabetic writing systems, such as the Phoenician alphabet around 1200 BCE, simplified writing and made it more accessible [337]

Classical Antiquity (500 BCE - 500 CE)
Flourishing of civilizations such as Greece and Rome, characterized by advancements in philosophy, science, and the arts. Development of democracy, philosophy, literature, and monumental architecture [339]

Middle Ages (500 - 1500 CE)
Feudalism, the rise of Christianity and Islam, and the establishment of universities. Advances in art (Gothic architecture), literature, and the preservation of knowledge through monasteries [340]

Renaissance to Enlightenment (14th - 18th centuries)
A revival of classical learning and a focus on humanism, science, and exploration [341,342].

The Printing Revolution
The invention of the printing press by Johannes Gutenberg in the 15th century revolutionized communication by making books more accessible and promoting literacy [343].

Art: The Renaissance saw the emergence of perspective in painting, with artists like Leonardo da Vinci and Michelangelo creating masterpieces that emphasized human emotion and realism.
Science: The Scientific Revolution introduced empirical methods and challenged traditional views, with figures like Galileo and Newton laying the groundwork for modern science.

Political Thought: Enlightenment thinkers such as John Locke, Voltaire, and Rousseau explored ideas of democracy, individual rights, and social contracts, influencing modern political systems.

Industrial Revolution (18th - 19th centuries)
A period of rapid industrialization, urbanization, and technological innovation [344,345].

Technology: The invention of the steam engine, mechanized textile production, and advancements in transportation (railroads, steamships) transformed economies and societies.
Social Changes: The rise of the working class, changes in labor practices, and the beginnings of labor rights movements.
Art and Literature: Movements such as Romanticism emerged, emphasizing emotion and nature in response to industrialization.

Modern Era (20th century - Present)
A period marked by global conflicts, technological advancements, and significant social changes [346,347].

Modern Communication
The 19th century saw the invention of the telegraph (1837) and the telephone (1876), which transformed long-distance communication [348] The 20th century introduced radio (1920s) and television (1930s), further changing how information was disseminated and consumed [349] The late 20th and early 21st centuries have seen the rise of the internet, social media, and instant messaging, which have transformed communication on a global scale [347].

The cultural evolution from early hominins to modern humans reflects a complex interplay of technological advancements, social structures, and symbolic thought. Each stage of development has contributed to the rich tapestry of human culture, shaping our societies, beliefs, and interactions today. Understanding this progression helps us appreciate the depth of human history and the factors that have influenced our current world [350,351].

8.3.4 Early civilization and early written Law and Religion
Two critical aspects of early civilizations were the development of law and religion, which played significant roles in maintaining order and

providing a framework for social cohesion. Critical to this process was written law and other governmental and religious writings.

One of the earliest and most significant legal codes was established by King Hammurabi of Babylon around 1754 BCE. The Code of Hammurabi consisted of 282 laws that covered various aspects of daily life, including trade, family relations, and property rights. It is notable for its principle of retributive justice, often summarized as "an eye for an eye." This code exemplified the move from arbitrary rule to a more structured legal framework, emphasizing the importance of written laws in governance [352]

In ancient Rome, the development of law was crucial for managing the complexities of a vast empire. The Twelve Tables (c. 450 BCE) were among the earliest attempts to codify Roman law, providing a public record of legal standards. Roman law evolved over centuries, influencing legal systems in many modern countries [353].

Religious Symbolism
Much of the development of symbolism was development of Religious Symbolism, here are some contexts that are foundational:
Polytheism: Early civilizations often practiced polytheism, worshipping multiple gods associated with natural elements and societal functions. In Mesopotamia, for example, deities like Anu (the sky god) and Enlil (the god of wind) were central to religious practices. Temples served as both religious and economic centers, reinforcing the connection between religion and governance [354].
Egyptian Religion: Ancient Egypt had a rich religious tradition centered around gods like Ra (the sun god) and Osiris (the god of the afterlife). The belief in an afterlife led to elaborate burial practices and monumental architecture, such as the pyramids. Religion was intertwined with the state, as pharaohs were considered divine figures [355].
Zoroastrianism: Emerging in ancient Persia, Zoroastrianism introduced concepts of dualism, with a focus on the struggle between good (Ahura Mazda) and evil (Angra Mainyu). This religion influenced later monotheistic faiths, including Judaism, Christianity, and Islam [356].

8.4 The Logos: 'word-centric' reality
In this section we briefly explore the implications or meaning in a 'word-centric' reality. To start, we see that word has meaning from language (e.g., a sequence of words with meaning in that language). In turn,

language is thought to shape consciousness by many (details to follow). Consciousness itself is viewed as a sequential process (initial details follow, then developed further in Ch. 9). There is also the notion that conscious reality is constructed through the narratives and vocabularies we use. This, in turn, has profound societal implications. So q quick sampling of these ideas is now given:

Ludwig Wittgenstein - *Philosophical Investigations*
In this seminal work, Wittgenstein challenges the idea that language is a fixed system of signs that corresponds directly to reality. Instead, he argues that the meaning of words is derived from their use in specific contexts (language games). This perspective emphasizes that our understanding of reality is mediated through language. "The meaning of a word is its use in the language." [357].

Ferdinand de Saussure - *Course in General Linguistics*
Saussure's work laid the foundation for modern linguistics and introduced the idea that language is a system of signs. He distinguished between the "signifier" (the word) and the "signified" (the concept), arguing that the relationship between them is arbitrary. This arbitrariness highlights how language shapes our perception of reality. "In language, there are only differences." [358].

George Lakoff and Mark Johnson - *Metaphors We Live By*
Lakoff and Johnson argue that metaphors are fundamental to our understanding of the world and that they shape our thoughts and actions. They propose that our conceptual system is largely metaphorical, and these metaphors influence how we perceive reality. "Our conceptual system is fundamentally metaphorical in nature." [359].

Mikhail Bakhtin - *The Dialogic Imagination*
Bakhtin's work emphasizes the social and dialogic nature of language. He argues that meaning is created through interaction and that language is a living force that reflects the complexities of social life. This perspective underscores the idea that our reality is constructed through the words we use in dialogue with others. "Language is not a neutral medium; it is a living, social force that shapes consciousness." [360]

Jacques Derrida - *Of Grammatology*
Derrida's deconstructionist approach challenges the traditional hierarchy of speech over writing, arguing that writing is not merely a secondary

representation of spoken language. He posits that language is inherently unstable and that meaning is always deferred, suggesting that our understanding of reality is contingent upon the play of words. "There is nothing outside of the text." [361]

Viktor Frankl - *Man's Search for Meaning*
Frankl's exploration of meaning emphasizes the importance of words and language in shaping human experience. He argues that individuals can find meaning even in the most difficult circumstances through the language they use to frame their experiences. [362]

Daniel Dennett - *Consciousness Explained*
Dennett explores the relationship between language, thought, and consciousness. He argues that language is a crucial component of human consciousness and that our understanding of reality is mediated through linguistic constructs. "Consciousness is not a thing, but a process." [363].

Richard Rorty - *Philosophy and the Mirror of Nature*
Rorty critiques the traditional view of philosophy as a reflection of reality, arguing instead that knowledge is a product of language and social practices. He suggests that instead of seeking objective truths, we should focus on the utility of language in achieving our goals and understanding our experiences. This perspective emphasizes that our reality is constructed through the narratives and vocabularies we adopt [364]

Edward Sapir - *Language: An Introduction to the Study of Speech*
Sapir, a linguist and anthropologist, explored the relationship between language and culture. He argued that language is not just a means of communication but also a way of shaping thought and perception. His work suggests that the structure of a language influences how its speakers conceptualize the world, reinforcing the idea of a word-centric reality. "Language is a guide to social reality." [365]

Benjamin Lee Whorf - *Language, Thought, and Reality*
Whorf is known for his hypothesis that language influences thought, often referred to as linguistic relativity or the Sapir-Whorf hypothesis. He argued that the structure of a language affects its speakers' worldview and cognition, suggesting that different languages lead to different ways of experiencing reality. "We dissect nature along lines laid down by our native languages." [366]

George Orwell - *1984*

In his dystopian novel, Orwell explores the concept of "Newspeak," a language designed to limit freedom of thought and expression. The manipulation of language serves as a tool for controlling reality and shaping consciousness, illustrating the profound impact that language can have on perception and ideology. "The purpose of Newspeak is to narrow the range of thought." [367].

To explore the Logos in the sense of 'Narrative Reality' further, we enter the realm of related material on consciousness, as will be described in the next chapter.

Chapter 9. Emanation and Emergence of Consciousness

9.1 Language, Narrative, and Consciousness

In Chapter 8 we saw that part of the mystery of the Logos reduced to the mystery of consciousness itself, and did so from a 'word-centric' foundation. This chapter picks up where Ch. 8 left off, with an exploration of word, language, and narrative as foundational elements of consciousness. Consciousness is probably more than a multilayered (sequential) narrative (Sections 9.1-9.4), but just considering the sequential nature of consciousness, where one narrative dominates, requires that there is a sequential nature to consciousness. Perspectives on this from different fields are given in Section 9.5. We can perceive consciousness in others and ourselves (to some extent, especially upon awakening). Awakening consciousness is mentioned in two contexts in Section 9.6 (individual and civilizational). Even as we begin to see glimpses of how consciousness might physically arise, the mechanism typically involves an infinite set of possibilities projecting or emanating to a finite set (or a smaller infinite set somehow). So we are connecting with infinity, in some sense, in formulating a satisfactory *physical* model for consciousness. Thus, an element of the super-natural, infinity, is still needed. This is contradictory to having a physical, not supernatural, description of consciousness. So, if consciousness requires the supernatural in its explanation, perhaps it's not just consciousness but also spirituality that must be discussed (with its obvious supernatural connections). In other words, perhaps 'consciousness' and 'spirituality' are inseparable and there is really the matter of 'consciousness/spirituality' to be explored? For this chapter we will proceed with consciousness where any infinity is approximated by a large number, and keep the description finite and leave the possible complications of dealing with infinity and the supernatural to Chapter 10.

<u>Quotes on Language and Consciousness</u>

Gottfried Wilhelm Leibniz (1646-1716): "We must consider what is the nature of the language of thought, and how it is that we can express our thoughts in words." [368]

Friedrich Nietzsche (1844-1900): "We have our thoughts, and we have our words; but we do not have our thoughts in our words. The words are merely the clothing of our thoughts." [369].

Ludwig Wittgenstein (1889-1951): "The limits of my language mean the limits of my world." [370]

Noam Chomsky (1928-present): "The study of language is a part of the study of the mind, and the mind is a part of the study of consciousness." [371]

Viktor Frankl (1905-1997): "Between stimulus and response there is a space. In that space is our power to choose our response. In our response lies our growth and our freedom." [362]

David Bohm (1917-1992): "Language is a very important part of the process of thought, and thought is a very important part of the process of consciousness." [372]

Mikhail Bakhtin (1895-1975): "Language is not a neutral medium; it is a living, social force that shapes consciousness." [360]

Howard Gardner (1943-present): "Language is a window into the mind, revealing the ways in which we think and perceive the world." [373]

Daniel Dennett (1942-present): "Consciousness is not a thing, but a process, and language is one of the most important processes that shape our conscious experience." [363].

9.2 Tarot-complete Narrative Generation

Tarot-based narrative generation is a creative process that utilizes the symbolism and meanings of tarot cards to inspire storytelling, character development, and plot creation. This method can be employed in various contexts, including writing, role-playing games, and personal reflection. Below is an overview of tarot-based narrative generation.

Overview of Tarot-Based Narrative Generation [374-393]

1. **Tarot Cards**: Tarot decks typically consist of 78 cards, divided into the Major Arcana (22 cards representing significant life

themes and archetypes) and the Minor Arcana (56 cards divided into four suits, representing everyday events and experiences).

2. **Symbolism and Meaning**: Each tarot card carries specific meanings, symbols, and associations that can be interpreted in various ways. These meanings can serve as prompts for narrative elements, such as character motivations, conflicts, and resolutions.
3. **Narrative Structure**: Tarot can help structure narratives by providing a framework for plot development. For example, drawing cards can suggest the beginning, middle, and end of a story, or highlight key turning points.
4. **Character Development**: Tarot cards can be used to develop characters by exploring their traits, desires, and challenges. Each card can represent different aspects of a character's personality or journey.
5. **Creative Prompts**: Writers and creators can use tarot readings as a source of inspiration, allowing the cards to guide their imagination and generate new ideas.
6. **Role-Playing Games**: In tabletop RPGs, tarot can be integrated into gameplay to influence story arcs, character decisions, and world-building.

9.3 Narrative with Scientific/Observational Data: The Book of Enoch [394]

The Book of Enoch, also known as 1 Enoch, is an ancient Jewish apocalyptic text attributed to Enoch, the great-grandfather of Noah (the narrative is even more ancient, tracing to ancient Akkad/Sumer/Babylon). It is not included in the canonical Bible but is considered significant in the study of early Jewish thought and Christian origins. The book is divided into several sections, with the most notable being:

1. **The Book of Watchers**: This section describes the fall of the Watchers, a group of angels who descended to Earth and took human wives, leading to the birth of the Nephilim, giants who caused chaos and violence. God sends the Flood as a judgment against humanity and the Watchers.
2. **The Book of Parables**: This part contains visions of the coming judgment and the establishment of God's kingdom, emphasizing themes of righteousness and the fate of the wicked.
3. **The Astronomical Book**: This section discusses the movements of heavenly bodies and the calendar, reflecting an interest in cosmology and the natural order.

4. **The Book of Dream Visions**: Enoch recounts two visions that symbolize the history of Israel, including the judgment of the wicked and the vindication of the righteous.
5. **The Epistle of Enoch**: This final section contains ethical teachings and warnings about the coming judgment.

9.3.1 Repeated story elements and refinements in later religious texts

The Book of Enoch shares several themes and story elements with later Babylonian and Hebrew texts:

1. **The Nephilim**: The concept of the Nephilim, the offspring of the Watchers and human women, is echoed in later Jewish texts and interpretations of Genesis 6:1-4. The idea of divine beings interacting with humans and the resulting consequences appears in various mythologies.
2. **Divine Judgment**: The theme of divine judgment, particularly through catastrophic events like the Flood, is prevalent in both Enoch and the Babylonian Epic of Gilgamesh, which also features a flood narrative (the story of Utnapishtim). Both texts emphasize the moral failings of humanity leading to divine retribution.
3. **Apocalyptic Visions**: The apocalyptic visions in Enoch, including the final judgment and the establishment of a new order, resonate with later Hebrew prophetic literature, such as the Book of Daniel and the Revelation in the New Testament.
4. **Angelic Hierarchies**: The depiction of angels and their roles in the divine order in Enoch influenced later Jewish and Christian angelology, as seen in texts like the Book of Revelation and various apocryphal writings.
5. **Ethical Teachings**: The moral and ethical teachings found in Enoch, particularly regarding righteousness and the consequences of sin, are echoed in later Jewish wisdom literature, such as Proverbs and Ecclesiastes.

9.3.2 Externalized collective unconscious

Overall, the Book of Enoch serves as a bridge between ancient Near Eastern mythology and later Jewish and Christian thought, with its themes and narratives influencing a wide range of religious texts and traditions. The Book of Enoch, in other words, is one of the earliest examples of a highly refined and recurring narrative. The repeated themes over thousands of years, in various forms, gives rise to an external (written) collective unconsciousness for the various civilizations carrying forward their version of the narrative. A similar process of internal (linguistic) collective unconsciousness is addressed in the next section, by Jung.

9.4 (Internalized) Collective Unconscious (Jung)

Carl Jung's theory of the collective unconscious posits that beneath the personal unconscious lies a deeper layer of the psyche that contains universal memories, symbols, and archetypes shared by all humanity.

Carl Jung (1875-1961), was a Swiss psychiatrist and psychoanalyst, who introduced the concept of the collective unconscious in his seminal work, *The Archetypes and the Collective Unconscious* (1959) [395]. Jung argued that the collective unconscious is a reservoir of experiences and knowledge inherited from our ancestors, containing archetypes—innate, universal symbols and themes that shape human behavior and thought.

Archetypes

Jung identified several key archetypes that manifest in myths, dreams, and cultural narratives. Some of the most prominent archetypes include:

1. **The Self:** Represents the unified consciousness and the integration of the personality.
2. **The Shadow:** Embodies the darker, repressed aspects of the psyche, often associated with fear and the unknown.
3. **The Anima/Animus:** Represents the feminine aspect in men (anima) and the masculine aspect in women (animus), symbolizing the balance of gender qualities within the psyche.
4. **The Hero:** Represents the quest for self-discovery and the journey toward individuation, often facing trials and challenges.
5. **The Wise Old Man/Woman:** Symbolizes wisdom, guidance, and the connection to the spiritual realm.

These archetypes are not only psychological constructs but also manifest in cultural myths, religious symbols, and artistic expressions across different societies, indicating a shared human experience. Jung believed that the collective unconscious plays a crucial role in shaping individual consciousness and behavior. He argued that by exploring the archetypes within the collective unconscious, individuals can gain insight into their own psyche and the broader human experience [78, 395-401].

9.5 Sequential Nature of Consciousness [78,402-405]

The concept of consciousness as a sequential process can be intriguing, especially when drawing parallels to systems like the tarot deck, which consists of 78 cards that can represent various narratives and archetypes. This analogy suggests that consciousness might similarly be structured around a finite set of mental states, each representing a different aspect of thought or experience.

Suppose consciousness consists of a sequential saccade over a 78 possible state sequence of mental states, where 78 is chosen since the 78-card tarot deck is supposed to generate all possible narratives, thus 78 different types of mental states might similarly underly what we regard as a 'train of thought' or a 'stream of consciousness'.

The term "stream of consciousness" was popularized by literary figures like Virginia Woolf and James Joyce, who depicted the continuous flow of thoughts and feelings in their characters. Psychologically, this concept aligns with William James's description of consciousness as a "stream" that is constantly changing and evolving. The linearity of this stream suggests that thoughts are not isolated but rather interconnected, forming a narrative or a sequence.

The analogy of a "saccade"—a rapid movement of the eye between fixation points—can be applied to how consciousness shifts focus between different mental states. Just as the eye quickly moves to gather information from various points in the visual field, consciousness may rapidly transition between different thoughts or states of awareness. This can be likened to the cognitive process of attention, where the mind selectively focuses on certain stimuli while ignoring others.

The 78-State Model

The idea of having 78 distinct mental states suggests a structured approach to understanding consciousness. Each state could represent a different emotional, cognitive, or perceptual experience. This finite model resonates with theories in cognitive psychology that propose categorization of thoughts and emotions, such as the basic emotions theory by Paul Ekman, which identifies a limited number of fundamental emotions that underpin human experience.

The tarot deck, with its archetypal imagery and symbolism, serves as a rich narrative framework. Each card can evoke specific themes, emotions, and insights, paralleling how different mental states can shape our understanding of experiences. Carl Jung's concept of archetypes aligns with this idea, suggesting that certain universal symbols and themes recur in human thought and culture, influencing our consciousness.

Note: The notion that mental states are interconnected can be explored through the lens of cognitive networks. Research in neuroscience indicates that thoughts and memories are stored in a network-like structure in the brain, where activation of one node (or thought) can lead to the activation of related nodes. This interconnectedness supports the idea that ***consciousness is not merely a linear sequence*** but a complex web of associations. The construction of narratives from these mental states is crucial for understanding human experience. According to narrative psychology, individuals create personal narratives that help them make sense of their lives. This process involves selecting and organizing experiences into a coherent story, much like how a tarot reading might unfold a narrative based on the cards drawn.

The analogy of consciousness as a sequential saccade over a 78-state model offers a compelling framework for understanding the complexity of human thought and experience. By integrating concepts from psychology, neuroscience, and narrative theory, we can appreciate the structured yet fluid nature of consciousness. This perspective not only highlights the interconnectedness of mental states but also emphasizes the importance of narrative in shaping our understanding of self and reality.

Sequential and Non-Linear Models [406]

The idea of consciousness as a sequence of mental states aligns with both psychological theories and philosophical inquiries. The linearity of a "train of thought" suggests a structured progression, yet consciousness often exhibits non-linear characteristics, where thoughts can loop back or diverge unexpectedly. This duality is reflected in the work of cognitive scientists like Daniel Kahneman, who distinguishes between two modes of thinking: the fast, intuitive System 1 and the slower, more deliberate System 2 [406]. This model implies that consciousness can oscillate between different states.

The Role of Attention [407]

Attention plays a crucial role in shaping consciousness. Theories of selective attention, such as the spotlight model, suggest that consciousness can be likened to a spotlight that illuminates certain mental states while leaving others in the dark [407]. This selective focus can be seen as analogous to the tarot's ability to highlight specific narratives or archetypes, guiding the individual's understanding of their experiences.

Archetypes and Symbolism

The tarot deck's 78 cards can be viewed as a representation of archetypal narratives, resonating with Jungian psychology. Jung posited that archetypes are innate, universal symbols that shape human experience and consciousness [78]. Each card in the tarot can evoke specific themes, emotions, and insights, paralleling how different mental states can influence our understanding of reality. This suggests that consciousness may be structured around a finite set of archetypal experiences, much like the tarot's narrative framework.

Narrative Psychology

The construction of personal narratives is essential for understanding consciousness. Narrative psychology posits that individuals create stories to make sense of their lives, integrating various experiences into a coherent whole [405]. This process mirrors the way tarot readings can unfold narratives based on the cards drawn, suggesting that consciousness is not merely a collection of isolated thoughts but a dynamic interplay of interconnected mental states.

9.5.1 Physics Perspective

9.5.1.1 Penrose – infinity from quantum aspect (via reduction process)

Penrose and Orchestrated Objective Reduction (Orch-OR)

Roger Penrose, a prominent physicist and mathematician, has proposed a theory of consciousness known as Orchestrated Objective Reduction (Orch-OR), developed in collaboration with anesthesiologist Stuart Hameroff. This theory posits that consciousness arises from quantum processes within neuronal microtubules [408,409]. According to Orch-OR, the collapse of quantum wave functions in these microtubules leads to moments of conscious experience, suggesting a fundamental link between consciousness and the fabric of the universe.

Integrated Information Theory

Penrose's ideas have sparked significant debate within both the physics and neuroscience communities. Some researchers argue that quantum mechanics may play a role in cognitive processes, while others remain skeptical, emphasizing the challenges of reconciling quantum phenomena with classical neuroscience. Critics argue that the quantum effects proposed by Penrose may not be significant enough to influence neural processes, given the warm, wet environment of the brain [410,411]. Alternative theories, such as Integrated Information Theory (IIT) proposed by Giulio Tononi [412], suggest that consciousness arises from the integration of information within a system, providing a different framework for understanding the nature of conscious experience.

9.5.1.2 Emanation – three infinities from emanation aspects

Each of the derivations of alpha described in Ch. 2 and App. A, refer to a from of infinity. The first derivation in volves the maximum perturbation allowed in evolving a doubly chiral trigintaduonion, where a maximal emanation is described from an infinite source. The second derivation involves maximal perturbation on an exponentiated map of the trigintaduonion. The third derivation derives alpha from pushing to the fractal limit at the "edge of chaos", where again an infinite (limit) process is involved.

9.5.2 Neuroscience Perspective

Recent advancements in neuroscience have provided insights into the neural correlates of consciousness. Studies using functional magnetic resonance imaging (fMRI) and electroencephalography (EEG) have identified specific brain regions and networks associated with conscious awareness [413]. These findings suggest that consciousness may emerge from the dynamic interactions of neural circuits rather than being localized to a specific area of the brain [413]

The Global Workspace Theory

One influential model in neuroscience is the Global Workspace Theory (GWT), proposed by Bernard Baars. GWT posits that consciousness arises from the broadcasting of information across a network of neurons, allowing for the integration of diverse cognitive processes. This theory aligns with the idea of a sequential saccade over mental states, as it emphasizes the dynamic nature of conscious experience [414]

9.5.3 Philosophical and Psychological Perspectives

Dualism vs. Physicalism: Philosophers like René Descartes proposed dualism, the idea that the mind and body are distinct entities. In contrast, physicalists argue that consciousness arises from physical processes in the brain.

Qualia [415,416]: This term refers to the subjective, qualitative aspects of conscious experience (e.g., the redness of red). Philosophers like

Thomas Nagel and Frank Jackson have explored the implications of qualia for understanding consciousness.
The Hard Problem of Consciousness [417]: Coined by David Chalmers, this term refers to the difficulty of explaining why and how subjective experiences arise from neural processes.
Cognitive Psychology [418]: This field studies the mental processes underlying consciousness, including perception, attention, and memory. Researchers investigate how these processes contribute to our conscious experience.
Altered States of Consciousness [419]: Psychologists explore various states of consciousness, such as dreaming, meditation, and drug-induced states, to understand how consciousness can change and what these changes reveal about the mind.

9.6 Awakening Consciousness

9.6.1 Sartre at the Beach [420] (Individual Logos)

Recall Jean-Paul Sartre's novel "Nausea" ("La Nausée"), where he vividly describes the experience of being at the beach and the overwhelming sensations that accompany it. Here is a well-known excerpt from that section:
"I was at the beach, and I was overwhelmed by the sensations of the world around me. The sun was shining, the waves were crashing, and I felt the weight of existence pressing down on me."

Original French Version [421]
"J'étais à la plage, et j'étais submergé par les sensations du monde qui m'entourait. Le soleil brillait, les vagues s'écrasaient, et je ressentais le poids de l'existence peser sur moi. Les couleurs étaient plus intenses, les sons plus vifs, et j'étais pleinement conscient de ma propre existence à ce moment-là."

9.6.2 Eisley (Civilizational Logos)

Loren Eisley's *The Immense Journey* [422-424], first published in 1957, is a profound exploration of the relationship between humanity and the natural world, as well as the evolution of consciousness. The book is a collection of essays that blend science, philosophy, and personal reflection, showcasing Eisley's deep appreciation for the mysteries of life and the universe.
Eisley, an anthropologist and naturalist, uses his poetic prose to convey the wonder of existence and the interconnectedness of all living beings. He reflects on the journey of evolution, emphasizing the significance of

human consciousness as a product of millions of years of development. The essays explore themes such as the origins of life, the nature of time, and the role of humanity in the cosmos.

One of the central ideas in *The Immense Journey* is the concept of the "immense journey" itself, which refers to the evolutionary path that has led to human awareness and self-reflection. Eisley contemplates the struggles and triumphs of life, the beauty of nature, and the existential questions that arise from our awareness of mortality. He invites readers to consider their place in the universe and the responsibilities that come with consciousness.

Chapter 10. Eucatastrophe and Human Spirituality

In this chapter we take the concept of consciousness further, to transcendent consciousness and spirituality. The order of presentation in the chapters thus far has shown how one emergent construct or phenomenon has built on what came prior. For the matter of consciousness and spirituality, however, this ordering and separability is not to be inferred from the separate Chapter 9 discussion on consciousness, and now the material in this chapter. In fact, an argument can be made (Occam's Razor: "Entities should not be multiplied without necessity" [425]) that consciousness and spirituality are not two things but one thing.

The overall subject of spirituality is outside the scope of this book. What is being attempted in this last chapter is to carry forward the ramifications of the hypothesis of maximal information emanation to even these lofty domains to see what, if any, 'high-level' correspondence might be discerned. This in not meant to be an in-depth analysis on any particular, so sections will be brief. What will be made clear, however, is that emanator theory profoundly links science and religion in unexpected ways, like last occurred during the Renaissance.

Eucatastrophe is a term coined by J.R.R. Tolkien to describe a sudden and favorable resolution of events in a story, particularly in a way that brings about joy and hope. In the context of human spirituality, a eucatastrophe can be seen in the existence of such spirituality.

10.1 Life is a kabbalistic optimum

Life can be understood through the lens of Kabbalah, a mystical Jewish tradition that emphasizes the interconnectedness of all things and the pursuit of spiritual enlightenment. The idea of an "optimum" refers to a state of balance and harmony in which individuals can achieve their highest potential. Kabbalistic concepts such as the Tree of Life and the Sefirot illustrate how divine attributes manifest in the world [426,427].

10.2 Man is in the image of God (alpha evolution)

The theological concept that humans are created in the image of God [428], has implications for human dignity, purpose, and spirituality [429].

The term "alpha evolution" refers to the idea that humanity is in a constant state of evolution, both physically and spiritually, moving toward a divine ideal. In the context of the preceding chapters, this is the evolution towards alpha communication and alpha consciousness.

10.3 CS Lewis, Chomsky – there is inherent human structure

C.S. Lewis and Noam Chomsky both argue for the existence of an innate structure within humans (from very different perspectives). Lewis, in his writings on morality and human nature, posits that there is a universal moral law that reflects a divine order [430]. Chomsky's theory of universal grammar suggests that the ability to acquire language is hardwired into the human brain, indicating an inherent structure to human cognition [431].

C.S. Lewis explored the concept of innate human nature extensively in his works, particularly in relation to morality, the existence of a universal moral law, and the nature of humanity in relation to God. In "Mere Christianity," Lewis discusses the idea of a universal moral law that exists within all humans, suggesting that there is an innate understanding of right and wrong [430]

In "The Abolition of Man," Lewis argues against the idea that moral values are subjective and emphasizes the existence of an objective moral order that is inherent to human nature: "The man who has a 'moral' experience is not merely a man who has a feeling of approval or disapproval; he is a man who has a sense of obligation. He feels that he ought to do something, and that he ought to do it for a reason that is not merely personal." [431]

10.4 CS Lewis provides an early Multiverse argument

C.S. Lewis discusses the concept of parallel universes and the implications of the truth of Christianity in his book *The Discarded Image* [433] . In this work, he explores the nature of reality and the idea that if the story of Jesus were true in any universe, it would hold true universally: "If Christianity is true, then it is true for all worlds. If it is true in one world, it is true in all worlds. If it is true in one universe, it is true in all universes." This quote encapsulates Lewis's view on the universality of truth, particularly in relation to the Christian narrative, suggesting that the truth of such a story transcends individual realities or universes

10.5 Emanation has Unitarian/Trinitarian structure

Have Trinitarian/"Logos" structure. The Trinitarian Process (John 1:1 makes explicit reference to trinity in terms of logos):

$$T_{base} \to T_{Em}(T_{base}) \quad and \quad \delta T = T_{Em}(T_{base}) - T_{base},$$

for elements:

(1) T_{base} is "the word" or the "logos":
In the beginning was the Word,
(2) $T_{Em}(T_{base})$ is the "emanation" of the word:
and the Word was with God,
(3) δT is the change in the word, the "meta logos":
and the Word was God.

10.6 Emanation and Harmony (Interconnectedness)

Here, the concept of emanation is linked to the idea of harmony found in Eastern religions, such as Hinduism and Buddhism. These traditions often emphasize the interconnectedness of all beings and the importance of achieving balance and harmony within oneself and with the universe. The notion of emanation suggests that all existence flows from a divine source, creating a harmonious tapestry of life [434,435].

10.7 Fractal emanation and transcendent consciousness

This section posits that the concept of emanation implies the existence of a transcendent consciousness that underlies and permeates all reality. This consciousness is often associated with the divine or the absolute, suggesting that human spirituality is a reflection of this higher state. (App C for further details.)

10.8 Maximal emanation narrative and the soul

Principle of Maximal Narrative Continuity (an extension of MIE from physical theory to the narrative/physical metaverse theory) means there is an 'afterlife', thus existence of the soul.

10.9 Emanation links Science and Religion – Renaissance 2.0

Lennox [436] declares no connection, fundamentally, between science (modern physics) and religion, but with emanator theory we not only have such a connection to religion, it is richly ecumenical.

God's Creation as a Word-Centric Narrative Universe

If creation is a word-centric narrative, it suggests that language and storytelling are fundamental to understanding existence. This aligns with the idea that the universe is imbued with meaning and purpose [532].

Innate Purpose and Cosmic Process

The idea of an innate purpose in the universe implies that humans are part of a larger cosmic narrative of growth and enlightenment. This perspective encourages a sense of interconnectedness and responsibility [533].

Logos as a Narrative Fractal Echo

The idea of the Logos as a narrative fractal echo implies that the fundamental principles of creation and existence are reflected in various scales and forms throughout the universe. This perspective aligns with the notion of "emanation," where all things originate from a singular source (Ein Sof in Kabbalistic thought) and manifest in diverse ways. The fractal nature of the Logos suggests that the patterns of creation, language, and meaning repeat and resonate across different levels of reality, from the microcosm of individual lives to the macrocosm of the universe itself [438,439].

A second Renaissance would be characterized by a resurgence of humanistic values, technological innovation, and interdisciplinary collaboration. Lennox and Steven Mayer might argue that this period would emphasize the integration of science and the humanities [436,437], and I would now add religion, as before, since it is evidently possible. Thus, Emanator Theory has answered the mystery of alpha, and in doing so has opened the door to a new Renaissance of Science, Humanities, and Religion.

Appendix

A. Emanator Theory Overview 1
B. Biology Information Processing 1
C. Narrative world: multiverses, magic, monadic travel 1

Appendix A. Emanator Theory Overview

A.1 Introduction

In quantum physics unitary propagation is a standard part of the description. Efforts to move to algebras to describe such propagation leads to formulations based on the normed division algebras (real, complex, quaternion, and octonion). In an effort to achieve maximal information propagation we relax the unitarity condition and show that multiplication (right) on a unit norm trigintaduonion base by a unit norm chiral trigintaduonions emanator results in a new unit norm product [2]. A path is comprised of repeated (right) multiplications. Each step of the 'emanation' arrived at is a multiplication by a chiral trigintaduonion. Use of methods from noise budget analysis, a constructive perturbation analysis, as well as analysis relating to maximal perturbation according to the Kato Rellich theorem, show that the chiral trigintaduonion with maximal perturbation has magnitude α, precisely the fine structure constant. A relation between α and π results. Suppose repeated achiral emanation steps can be described as an iterative mapping, with unit-norm constraint resulting in a quadratic relation on components, we then expect the Feigenbaum universal bifurcation parameter, C_∞, to appear according to the number of independent dimensions in a chiral trigintaduonion emanation step and the precise form of the "emanator" construction. The number of effective dimensions is shown to be 29 plus a little more, and a relation between α, π and C_∞results that is in agreement with the choice of emanator examined in computational studies shown here. The computational studies with the emanator have also been explored via "random walks" in the trigintaduonion space during emanation and to explore noise additivity effects. Component-level evolution is seen to behave like a random walk, with random walk asymptotics (established computationally). This helps to establish that the Emanation process is Martingale, since random walk processes are Martingale.

Just from the propagation structure on one path we already see core emergent structure that results in a universal emanation with structural parameters 10,22,78,137 and perturbation maximum α=~1/137. The central notion in the universal emanation hypothesis is that there should be *maximal information flow*, where this is accomplished by finding the highest theoretical dimensionality of unit-norm 'propagation', here called an emanation, which turns out to be 10, then add the maximal perturbation that still allows unit-norm propagation, where that perturbation is into the space the 10D motion is embedded in, here a 32 dimensional (trigintaduonion algebra) space.

The existing Standard Model, and reasonable extensions for the massive neutrino, cannot explain the parameters of the model. Why they are 19, or so, in number, and why the local gauge structure exists with the odd-looking product form: $U(1) \otimes SU(2)_L \otimes SU(3)$.

At the heart of the quantum formulation underlying the standard model is the fundamental theoretical element known as the quantum propagator (corresponds to a complex unitary matrix). Most notably, if the wavefunction has unit norm at the outset, after unitary propagation it remains unit norm.

Efforts to generalize the quantum formulation by considering hypercomplex but non-unitary matrices have been stymied for a variety of reasons [45]. Given the existing close agreement of experimental observation and quantum predictions, it is hard to work within the unitary propagator-type theoretical framework and arrive at anything other than an equivalent quantum formulation. Instead of working within the existing theory, in Emanator Theory the objective is to obtain a generalized theory that can project the quantized Lagrangian and Standard Model (or closely related extended Standard Model). To seek a larger theory that projects the existing theory is an excellent way to break free of the Godel-Incompleteness Trap that occurs when working within the existing standard physics and standard model. The drawback, of course, is it's likely to be a far-fetched idea, whatever it is, so it will face a higher bar to be seen as even interesting, not to mention valid. In what follows a synopsis of emanator theory will be given, as well as a few of the latest results, that will hopefully be convincing on both accounts.

If trying to generalize to a theory that can project to a renormalizable quantum field theory (with the Standard Model parameters), there is the issue of why it should project the particular formulation described. Of all the possible projections, how do you argue why it should be the unitary propagator-type physics seen? Here the aforementioned difficulty in generalizing the existing propagator theory (from the inside out) is a strength as it explains why the projection should be the standard physics and standard model as seen. That being the case, there is still the matter of finding the missing structure elsewhere (the Path Integral Action-Lagrangian formulation itself, for example, and the Standard Model parameters) and then projecting that as the renormalizable quantum field theory with Standard Model parameters as seen experimentally.

A.1.1 Dirac used Lorentz Invariance

If trying to guess the mathematical basis of a generalized quantum theory, that would project the existing Lorentz Invariant quantum field theory, then Dirac provides a powerful lesson. Recall that Dirac's guess and derivation for the relativistic wave equation was purely based on seeking a representation that was Lorentz invariant. In doing this, Dirac discovered the Dirac Equation that describes spin-1/2 fermionic matter (all the fundamental particles).

In Dirac's approach 4-vectors, V^a, were used to describe the spinors (spin ½ fermions), where the length of the 4-vector is constant under Lorentz Transformation L:

$$L\left\{\frac{1}{2}\eta_{ab}V^aV^b\right\} = \frac{1}{2}\eta_{ab}V^aV^b.$$

In Penrose's book on spinors [440] we see how to write a 4-vector as a 2x2 Hermitian matrix, $\psi(V^a)$, where the length of the four vector is equal to the determinant of the equivalent 2x2 matrix:

$$det[\psi(V^a)] = \frac{1}{2}\eta_{ab}V^aV^b.$$

Suppose the length of the 4-Vector is 1 (it's a spinor probability amplitude), then the associated 2x2 Hermitian matrix will have determinant equal to 1, thus $SL(2,\mathbb{C})$, and we have a direct representation of the Lorentz Group.

Let's now consider a similar process involving transformational invariance but instead of encoding the Lorentz transform in the form of matrix transformation invariance let's use elements of the Cayley algebras instead. Specifically, consider the following transformation:

$$q' = aqa_c^*, \ \ where \ \ aa_c = 1,$$

where a is a (unitary) complex bi-quaternion: $H(\mathbb{C}) \times H(\mathbb{C})$, and $q = (ct, ix, iy, iz)$ (note this notation is for $q = (ReH_1, iImH_1, iReH_2, iImH_2)$). The $q' = (ct', ix', iy', iz')$ that results will correspond to a proper orthochronous Lorentz transform [441-443].

This is a remarkable result, but is there better (higher dimensionality/complexity)? Consider that a complex bi-quaternion is isomorphic to a complex octonion which is isomorphic to an 'achiral' sedenion, thus a theory for achiral sedenion emanation is indicated from this result as far back as 1917. The 'halting condition' on the generalization in 1917 seems to be that octonions are the highest-order of the division algebras that have inverses defined (which is necessary to have $aa_c = 1$ be defined). But they have already extended past octonions since these are complex octonions ≅ sedenions, so how are they guaranteed to have $aa_c = 1$ be defined? This is possible for the *chiral* sedenions, as shown in [2], if restricted to be unit norm. So now we have our answer based on the results from [2] – we can go one complexation order higher:

$$Q' = AQA_c^*, \ where \ \ AA_c = 1,$$

where A is a (unitary) bi-complex bi-quaternion: $H(\mathbb{C} \times \mathbb{C}) \times H(\mathbb{C} \times \mathbb{C})$, which is isomorphic to a unit norm quaternionic bi-quaternion, $H(\mathbb{H}) \times H(\mathbb{H})$, and $Q = (ct, ix, iy, iz)$.
where $Q = (ReH_1, iImH_1, iReH_2, iImH_2)$ as before, except $H_1 = H \times \mathbb{H}$ not $H_1 = H \times \mathbb{C}$
The $Q' = (ct', ix', iy', iz')$ that results will again correspond to a proper orthochronous Lorentz transform. Again, how do we know the critical operation $AA_c = 1$ can always be satisfied? Previously we saw that a complex bi-quaternion was equivalent to a 'chiral' sedenion (in the sense described in [2]), thus a bi-complex bi-quaternion is isomorphic to a complex chiral sedenion, which is isomorphic to a 'doubly chiral' trigintaduonion. This is precisely the construct examined in [2], so a generalization of the 1917 result to unitary quaternionic bi-quaternions appears possible. As shown in [2], however, there is no higher order construct. This latter form (on doubly-chiral trigintaduonions) establishes the Emanator as Lorentz Invariant.

Thus, in terms of Cayley algebras, we can generalize to encoding the Lorentz transformation into split-Cayley algebras as long as the Cayley transform has an inverse, and doing this for the highest order Cayley algebra possible. The Cayley algebras with inverses, the division algebras, consist of the first four Cayley Algebras: reals (1 parameter), complex numbers (2 parameter), quaternions (4 parameters), and octonions (8 parameters). Another aspect of the division algebras is that they have norm, and this then means that their unitary evolution can be described as unit norm propagation. Suppose we focus on this latter feature, unit norm propagation alone. Can we extend to even higher order Cayley Algebras in the sense of starting with unit norm and for a subset of that higher order algebra still effect a unit norm, invertible, propagation? In other words for the next higher algebras, sedenions (16 parameter) and trigintaduonion (32 parameter), etc., does there exist a subspace allowing unit norm propagation? This is equivalent to considering the maximal Cayley subalgebra order for which unit norms exist. This is the actual starting point of the Maximal Information Emanation Hypothesis

A.1.2 The Maximum Information Emanation (MIE) Hypothesis and Prior Results

Emanator theory stems from the Maximum Information Emanation (MIE) Hypothesis. The definition of information is context dependent, so how the MIE hypothesis will manifest depends on circumstance. We start with the fundamental notion of the quantum propagator, for which mathematical 'propagators' satisfy unitarity. We seek to extend this foundational element so start by asking what is the highest Cayley algebra that can remain unitary – where the answer comes down to the highest order division algebra, which are the octonions. What if we change the desired property from unitary to unit-norm preserving? Then we can extend 2 more dimensions beyond the 8D octonion algebra to a chiral subspace of the trigintaduonions that is 10D [2].

The 10D chiral trigintaduonions are then identified as the maximal information 'carriers' or emanators, operationally like the quantum propagators in a larger theory, where evolution of the system will be shown to result from sums on paths of emanators (similar to quantum evolution in terms of a path integral on propagators 'steps'). Having posited this maximal construct we see the classic signs of "asking the right question" since we get a variety of clear results:

(1) the chiral emanator is manifestly Lorentz Invariant as indicated above, and the emanation step or propagation is simply multiplication by a unit-norm 'emanator' according to the Cayley algebra multiplication rules.

(2) chiral emanation involves a 10D element in a 32D space (trigintaduonions) for which maximum perturbation is determined computationally (still permitting unit-norm transmission) to be ~1/137, e.g., 'alpha' – we therefore have the mysterious alpha by a computational definition.

(3) The emanation process involves multiplication on the current unit norm trigintaduonion base element by a unit norm chiral trigintaduonion

emanator to arrive at a new, unit norm, trigintaduonion base (and then it repeats infinitely). When the trigintaduonion multiplication steps are expanded to octonionic level, we find that there are 137 independent tri-octonionic terms that occur in the emanation process. Due to complex noise contributions, the effective number is 137*, which is slightly greater than 137. Thus, theoretically, the maximum perturbation that is allowed is 1/137* which happens to be exactly 'alpha'. We, thus, have a theoretical derivation for alpha, referred to as the $\{\alpha, \pi\}$ relation:

$$\alpha^{-1} = \frac{137}{\cos\beta}\cos\theta\frac{\theta}{\sin\theta},$$

where

$$\beta = \frac{\pi}{137} \ and\ \theta = \frac{\pi}{137x29}$$

and

$$\alpha^{-1} = 137.03599978669910\,,$$

which agrees with 2002 experimental observation:

$$\alpha^{-1} = 137.03599976(50).$$

(4) Chiral trigintaduonion (32D) emanation with perturbation does not have effective dimension 32D due to chiral and other constraints -- noise budget analysis or (equivalently) Kato Rellich operator analysis, both indicate effective dimension slightly greater than 29 referred to as "29*". It is hypothesized that the maximal level of information flow dimensionally should relate to the Universal fractal constant C_∞ according to:

$$\alpha^{-1} = (C_\infty)^\gamma, \ \gamma \equiv 29^*,$$

where $\gamma$is estimated by [7]:

$$\gamma \cong \frac{1}{2}\left(29 + \left(\frac{4\pi}{72}\right)\left[1 + \left(\frac{\pi}{137 \times 29}\right)\left\{\frac{\pi}{72} + \frac{3}{72}\right\}\right]\right)$$

thus

$$\alpha^{-1} \cong 137.035999206 \ldots$$

in agreement with the exact alpha with nine digit accuracy.

In the process of estimating 29* we must explore the definition of the emanation process in the sense of is it one step then normalization, or a chain of steps (analogous to hands greater than one in size being dealt) then normalization, with infinite repeat. The answer appears to be two steps (since this suffices to 'flood' or max-out the noise channels, so no further steps needed) and working within this construct we obtain the $\gamma \cong 29^*$ estimate mentioned above. It's not an exact match to 16 decimal places and known measurement, but at a 9-decimal place match its pretty good, certainly indicating that there may well be a relation $\alpha^{-1} = (C_\infty)^{29^*}$ as hypothesized (sometimes called the $\{\alpha, \pi, C_\infty\}$ relation). And, if there is such a relation, it would indication that not only is evolution at maximum perturbation in the quantum sense, but it is also evolution at the edge-of-chaos in the thermal/statistical mechanical sense.

(5) the chiral emanator indicates 'motion' in a 10D subspace of 32D, suggesting 22 constants of the motion. This is a naïve analysis, but it turns out to be true upon deeper analysis within the emanator formalism as there are 22 types of emanation that result in no change to the base trigintaduonion describing the system. This, in turn, shows that the emanator theory will have 22 fixed parameters.

(6) By using the split form of the trigintaduonions, we not only have manifest Lorentz invariance, we also have an exact algebraic split to a space that is simply the direct product of 29* real dimensions (not a local approximation to such). This is important because the fundamental existence of a complex structure means that we trivially have the extension $\mathbb{R}^{29^*} \rightarrow \mathbb{C}^{29^*}$. Suppose we have point-like singular elements in $\mathbb{C}^{29^*}$, such will occur due to zero-divisors (zd's) in the 32D trigintaduonion space. To achieve a maximal domain of analyticity (an application of MIE), we must remove the zd-singularities. In doing so we obtain point-like matter in the theory and a small-h constant that enters the sum on emanator paths just as Planck's constant in the sum on propagator paths in the quantum formulation – suggesting that these small-h numbers are related.

(7) Three derivations of alpha are thus obtained: (i) Computational: $\{\alpha\}$ based on the maximal perturbation for which chiral emanation retains the unit-norm property; (ii) Theoretical: $\{\alpha, \pi\}$ based on the maximal noise transmission on a chiral emanation path; and (iii) Approximate: $\{\alpha, \pi, C_\infty\}$ based on the maximal noise transmission on an achiral emanation path (where maximal emanation is at "the edge of chaos" which is defined according to Feigenbaum Universality [41]).

(8) At component level in the emanation product, using 100's of millions of computational steps, we see an excellent asymptotic fit to random walk behavior. Since a random walk is a Martingale process, this strongly suggests that the achiral emanation process is Martingale (between normalization steps involved with zero crossings in their values). In turn,

the projected quantum process (standard theory) would retain the imprint of that Martingale process.

(9) The achiral emanator, a sum of achiral emanation paths, can be shown to have the mathematical form $\sum \exp(i\mathbb{H} \times \mathbb{O}) \rightarrow \mathbb{C} \times \mathbb{H} \times \mathbb{O}$, which can be shown to give the gauge theory of the standard model: $U(1) \times SU(2)_L \times SU(3)$. Thus, there is no grand unified theory in terms of gauges that is fundamental (although a GUT may approximately occur at early times cosmologically, at high temperature, where there is conformal flatness). The 'ugly' product gauge that is observed is precisely what is predicted by emanator theory.

(10) Universal thermality is indicated by application of the MIE hypothesis to the choice of whether 'effective' achiral emanation is associative or not, at least at the sedenion-level of propagation. In effect, the mechanism to extend unit norm (chiral) propagation on a 10D subspace of the 32D trigintaduonions can be repeated for other mathematical properties from the lower-level Cayley algebras. Thus, the mechanism identified for extending 'nice' properties of lower-order Cayley algebras to higher order (no more than two orders higher to be precise), can be used for other than existence of a norm. By the chiral extension mechanism, associativity can be extended to the chiral octonions and chiral sedenions. This provides the basis for the universal thermality relation that appears to be ubiquitous (e.g., analytic time). It also provides the basis for an associative matching of terms for cancellation, e.g., renormalization. Thus provides the mechanism to correct the projected quantum field theory such that it has the renormalization counter-terms needed.

(11) In the emanation process there is a clear separation between spinorial elements and manifold elements. Manifold elements include geometry and thermality, have no alpha-perturbation effects, and appear to be part of the 'apparatus' from the perspective of the quantum theory.

A.2 Synopsis of Methods and Prior Results

A.2.1 The Cayley Algebras

The list representation for hypercomplex numbers will make things clearer in what follows so will be introduced here for the first seven Cayley algebras:

Reals: $X_0 \rightarrow (X_0)$.
Complex: $(X_0 + X_1\, \boldsymbol{i}) \rightarrow (X_0, X_1)$ with one imaginary number.
Quaternions: $(X_0 + X_1\, \boldsymbol{i} + X_2 \boldsymbol{j} + X_3\, \boldsymbol{k}) \rightarrow (X_0, X_1, X_2, X_3)$ with three imaginary numbers.
Octonions: $(X_0, \ldots, X_7)$ with seven imaginary numbers.
Sedenions: $(X_0, \ldots, X_{15})$ with fifteen imaginary numbers.
Trigintaduonions (Bi-Sedenions): $(X_0, \ldots, X_{31})$ with 31 imaginary numbers.
Bi-Trigintaduonions: $(X_0, \ldots, X_{63})$ with 63 types of imaginary number.

Consider how the familiar complex numbers can be generated from two real numbers with the introduction of a single imaginary number '***i***', $\{X_0, X_1\} \rightarrow (X_0 + X_1\, \boldsymbol{i})$. This construction process can be iterated, using two complex numbers, $\{Z_0, Z_1\}$, and a new imaginary number '***j***':

$$(Z_0 + Z_1\, \boldsymbol{j}) = (\mathrm{A}+\mathrm{B}\boldsymbol{i}) + (\mathrm{C}+\mathrm{D}\boldsymbol{i})\, \boldsymbol{j} = \mathrm{A}+\mathrm{B}\boldsymbol{i} + \mathrm{C}\boldsymbol{j} + \mathrm{D}\boldsymbol{ij} = \mathrm{A}+\mathrm{B}\boldsymbol{i} + \mathrm{C}\boldsymbol{j} + \mathrm{D}\boldsymbol{k},$$

where we have introduced a third imaginary number '***k***' where '***ij=k***'. In list notation this appears as the simple rule ((A,B),(C,D)) = (A,B,C,D). This iterative construction process can be repeated, generating algebras doubling in dimensionality at each iteration, to generate the 1,2,4,8,16, 32, and 64 dimensional algebras listed above. The process continues indefinitely to higher orders beyond that, doubling in dimension at each iteration, but we will see that the main algebras of interest for physics are those with dimension 1,2,4,and 8, and sub-spaces of those with dimension 16 and 32 dimensional algebras.

Addition of hypercomplex numbers is done component-wise, so is straightforward. For hypercomplex multiplication, list notation makes the freedom for group splittings more apparent, where any hypercomplex product ZxQ to be expressed as (U,V)x(R,S) by splitting Z=(U,V) and Q=(R,S). This is important because the product rule, generalized by Cayley, uses the splitting capability. The Cayley algebra multiplication rule is:

$$\mathbf{(A,B)(C,D) = ([AC-D^*B],[BC^*+DA])},$$

where conjugation of a hypercomplex number flips the signs of all of its imaginary components:

$$(\mathrm{A,B})^* = \mathrm{Conj(A,B)} = (\mathrm{A}^*,-\mathrm{B})$$

The specification of new algebras, with addition and multiplication rules as indicated by the constructive process above, is known as the Cayley-

Dickson construction, and this gives rise to what is referred to as the Cayley algebras in what follows.

If a Split Cayley algebra is used, then the multiplication rule has a single sign difference:

$$\mathbf{(A,B)(C,D) = ([AC+D^*B],[BC^*+DA]).}$$

A.2.2 Unit-norm propagation

For a physical system, a unit norm object can be used to represent a system, and by repeated transformation to other unit norm objects, it thereby evolves. Mathematical objects that can effect this 'transformation' simply by the rule of multiplication would be objects like division algebras, ideals, and what I'll simply call projections or emanations. In the universal propagator we have a unit norm trigintaduonion (32D) and perform a right multiplication with a chiral (10D) unit norm 'alpha-step' (defined by a max perturbation α into the 29 free dimensions, given by 32 minus one for each chiral choice, and one for the unit normalization overall). Consider multiplication of a given (starting) trigintaduonion from the right with a chiral bi-sedenion as a 'projection' through the (chiral) step indicated. The repeated application and repeated 'chiral steps' thereby arriving at a path describing a chiral propagation. The resulting universal propagation consists of a 32D unit norm trigintaduonion with propagation via right multiplication using a unit-norm, chiral bi-sedenion, with max-α perturbation.

We thereby arrive at a 'Universe Propagator' that takes on the physics parameters desired (notably the fine-structure constant) and imprints them onto the evolution as seen from the 'internal reference frame' where we reference an object in the 4D spacetime with Standard Model gauge field, and where the standard Lagrangian emerges as the necessary 'propagate-able' structure (where Hilbert space must be complex, not real, quaternionic or octonionic, etc. [45]). From maximum information flow with the constructs, and the required emergent complex Hilbert space (thus complex path integral, thus standard quantum operator formalism) we arrive back at the familiar results with justification of their core mathematical representations (e.g., complex Hilbert space), and now with justification of all parameters, all from the emanation hypothesis.

Emanator Theory uses unit-norm propagation

Unit-norm right product propagation is trivial for the division algebras since norm(XY) = norm(X) × norm(Y). From this it is apparent that we have an automorphism group given by the norm itself (since an automorphism if A(XY)=A(X)A(Y)), and in the case of the octonions this automorphism group is G2 [46]. It can be shown that SU(3) is in G2. Let's now consider the situation with a higher-order Cayley algebra, the Sedenions, 'S'. We obviously don't have norm(S_1S_2) = norm(S_1) × norm(S_2) in general, as this would then allow S to join the ranks of the division algebras, and it is proven that such don't exist above the Octonions [47]. Can we still have a propagation structure? Is it possible to have a 'base' sedenion for which norm(S_{base})=1, and to have a right propagator (product) sedenion also norm(S_{right})=1, such that norm(S_{base} x S_{right}) =1? The answer is yes (see [126]), when the sedenion has the (chiral) form of an octonion crossed with a real octonion: S_{chiral} = (O,O_{real}) or S_{chiral} = (O_{real},O). Can we continue this to arrive at a propagation structure on the Trigintaduonions? Again the answer is yes, with the chiral form generalizing off the chiral Sedenion as might be expected: T_{chiral} = (S_{chiral}, S_{real}) or (S_{real}, S_{chiral}) [2]. It is proven that this extension process will go no further [2]. What happens is that due to the chiral form we are still able to re-express all T products (or S) as collections of terms involving tri-octonionic products (which have nice properties as described in [2]), and this can no longer occur above the (chiral) trigintaduonion level.

Thus, we have achiral emanation, and to get achiral there must be a way to sum over all chiral to get an achiral result to arrive at the full emanator process. These details are described next.

A.2.3 Chiral T-emanation has 78 generators of change and 137 independent octonion terms

We begin with constructing the theoretical expression for a general element of the trigintaduonion algebra after two chiral trigintaduonion multiplicative propagation steps. A simple analysis of the number of terms in this expression, when reduced to three-element algebraic 'braid-level', results in a count on algebraic braids of 137, plus a little extra (e.g. some lagniappe for the best 'cooking') of a contribution towards a 138th braid when the "noise analysis" is done [1,2].

Consider a general Norm=1 (32D) Trigintaduonion (Bi-Sedenion): (A,B), where A and B are sedenions (16D). Then have (A,B) = ((a,b), (c,d)), where {a,b,c,d} are octonions. Slightly different than a propagator, we have an 'emanator' with the following notation and properties, where the emanator describes a 10D multiplicative step. The emanator is a chiral bi-sedenion: a trigintaduonion whose first sedenion half is itself a chiral bi-octonion, and the second sedenion half is a pure real (as is the second octonion half): $(\tilde{A},\beta)$, $\tilde{A} = (\tilde{a},\alpha)$, where the norm is 1, α is a real octonion, and β is a real sedenion. Thus:

Emanator: $(\tilde{A},\beta) = ((\tilde{a},\alpha), \beta)$.
Note: $\tilde{A}^* = (\tilde{a}^*,-\alpha)$.

Let's set up a description of the Universal 'Emanation' along a 'chiral path' resulting from a few emanation steps. To begin, suppose we have already arrived at, or received, a unit norm trigintaduonion (32D) state 'T', and suppose our emanations are the result of right multiplication with a chiral trigintaduonion (bi-sedenion) 'step', and suppose we consider one such path after just a few steps. Here's the notation to begin:

$\mathbf{T} = (A,B)$, a unit norm trigintaduonion.
$\boldsymbol{\tau} = (\tilde{A},\beta) = ((\tilde{a},\alpha), \beta)$, the 'emanator' above (so named to distinguish from a 'propagator').

Universal Emanation from T on single path with three steps: $((\mathbf{T} \bullet \boldsymbol{\tau}_1) \bullet \boldsymbol{\tau}_2) \bullet \boldsymbol{\tau}_3)$ …

Consider the first emanation step:
$\mathbf{T} \bullet \boldsymbol{\tau}_1 = (A,B) \bullet (\tilde{A},\beta) = ([A\bullet\tilde{A}-\beta^*\bullet B] , [B\bullet\tilde{A}^*+\beta\bullet A])$. (Standard Cayley algebra multiplication rules.)
$A\bullet\tilde{A} = (a,b) \bullet (\tilde{a},\alpha) = ([a\bullet\tilde{a}-\alpha^*\bullet b] , [b\bullet\tilde{a}^*+\alpha\bullet a])$
$B\bullet\tilde{A}^* = (c,d) \bullet (\tilde{a}^*,-\alpha) = ([c\bullet\tilde{a}^*+\alpha^*\bullet d] , [d\bullet\tilde{a}-\alpha\bullet c])$
Thus,
$\mathbf{T} \bullet \boldsymbol{\tau}_1 = (A,B) \bullet (\tilde{A},\beta) = ([(a\bullet\tilde{a}-\alpha^*\bullet b-\beta c) , (b\bullet\tilde{a}^*+\alpha\bullet a-\beta d)] , [(c\bullet\tilde{a}^*+\alpha^*\bullet d+\beta a) , (d\bullet\tilde{a}-\alpha\bullet c+\beta b)])$.

At the lowest octonion level, that covers the pure real trigintaduonion, we have:

$(a\bullet\tilde{a}-\alpha^*\bullet b-\beta c) \rightarrow 8\text{x}8 + 8 + 8 - 2 = 64+14 = 78$ independent octonion terms (78 independent generators of motion). The –2 comes from the unit norm constraints on T and τ.

Now consider the second propagation step:

$\mathbf{(T \bullet \tau_1) \bullet \tau_2}$ = ([(a•$\tilde{a}$–α*•b–βc) , (b•$\tilde{a}$*+α•a–βd)] , [(c•$\tilde{a}$*+α*•d+βa) , (d•$\tilde{a}$–α•c+βb)]) • ($\tilde{A}$,β),
where $\boldsymbol{\tau_2}$ = ($\tilde{A}'$,β′) = (($\tilde{a}'$,α′), β′).

Let $\mathbf{(T \bullet \tau_1) \bullet \tau_2}$ = ([Z_{11},Z_{12}] , [Z_{21}, Z_{22}]).
$Z_{11} = (a\bullet\tilde{a}-b\alpha-c\beta)\bullet\tilde{a}' - (b\bullet\tilde{a}^*+\alpha a-\beta d)\ \alpha' - (c\bullet\tilde{a}^*+d\alpha+a\beta)\beta'$.

In Z_{11} we can replace the octonions with their unit component forms:

$$a = a_1e_1 + a_2e_2 + \ldots + a_8e_8\ ,$$

where {e_1, e_2, …, e_8} are the unit octonions (one real, seven imaginary), while 'α'=αe_9 and 'β'=βe_{17}, originally, but in expressions, are reduced to just their real part. All expressions, thus, involve 10 components: {e_1, e_2, …, e_8, e_9, e_{17}}, and as the equations for Z_{11} shows, grouped in factors of three (three-element octonionic 'braids'). We don't have associativity but we do have alternativity and the braid rules on three-element octonionic products that allows their regrouping. Applying these rules to have only ordered $e_i\bullet e_j\bullet e_k$ products in a simplified expression, we will then have 10x9x8/3! = 120 independent terms when the products involve different components. We have 8 independent terms when the first product are on the same component (equals 1), have 8 independent terms when the second product involves the same component, and have 1 independent term when the three-way product equals 1 (further details on this and the properties of the exponentiation map on hypercomplex numbers is given in the next section. There are, thus, 137 independent terms in Z_{11}, where each term has norm less than unity (since each octonionic component has norm less than one and the norm of a product of octonions is the product of their norms). The terms involving products with the same component, or with the components three-way product equal unity, correspond to the 'telescoping terms' in what follows.

When **T=((a,b),(c,d))** → $\mathbf{((T \bullet \tau_1) \bullet \tau_2)}$=(($Z_{11}$,$Z_{12}$),($Z_{21}$, Z_{22})). we have a→Z_{11} and the terms involving 'a' in Z_{11} are referred to as 'telescoping' due to their simple math properties with further emanation steps. In particular, the terms involving 'a' are:

Z_{11}[a terms]= $a\bullet\tilde{a}\bullet\tilde{a}' - a\alpha\alpha' - a\beta\beta'$.

We can see that the original 'a' information is passed along three (telescoping) channels, one involving repeated full octonionic factors $\tilde{a}$, one involving repeated real-octonion α factors, and one involving repeated real-octonion β factors:

(1) a → $(a \bullet \tilde{a}) \bullet \tilde{a}'$, if this product is continued indefinitely, then we have ***the random product of a collection of octonions***, all of which have norm less than one (although their norms can be quite close to one). If their norms were perfectly equal to one, then the addition of their random 'phases' would tend to cancel to zero, giving only a real octonionic component (same argument for phase cancelation on S1 as on S7 or S15). What results is a 'mostly' real octonion, having some imaginary part. A more precise, and lengthy, derivation is given in the next section.

(2) a → $a\alpha\alpha'$, if this product is continued indefinitely, 'telescoped' with repeated α products, we see that the original 8 independent terms arising from 'a' are passed forward with an overall real octonion product, giving rise to 8 independent terms.

(3) a → $a\beta\beta'$, as with (2), we have 8 independent terms.

From the above, we see an alternative accounting of the extra 17 independent terms to go with the 120 for a total of 137 independent terms in the propagation of the octonionic sectors of the universal emanation. A benefit of the telescoping analysis is it clarifies how in (1) an imaginary component may arise, and in perturbation expansions it will then be natural to refer to an overall imaginary component.

There are 137 terms in the dually chiral 'emanation', each with norm bounded by unity, with total bi-sedenion norm equal to unity. In the analysis that led to the computational discovery of α [126], an imaginary (non 10D) component was added of growing magnitude until unit-norm propagation failed. In essence, a maximum perturbation, from propagation strictly in the 10D subspace of the 32D trigintaduonions, was sought.

We identify maximal perturbation by doing an independent term analysis, and by adding a maximum perturbation term that implicitly identifies a definition of maximum antiphase. From this definition of maximum antiphase, there results the parameter π.

A.2.3.1 Exponential Map Properties when using hypercomplex numbers

For what follows, it helps to recall some important properties of the exponential, particularly its well-defined properties with hypercomplex numbers [1]. Important map relations:

(1) exponential map on Im(T) gives unit norm object: $\exp(\mathrm{Im}(T)\theta) = \cos\theta + \mathrm{Im}(T)\sin\theta$.

(2) exponential map on iT gives $C \times T$:
$\exp(iT) = \exp(i\mathrm{Re}(T)) \times \exp(i\mathrm{Im}(T)) = (\cos\theta + i\mathrm{Re}(T)\sin\theta) \times (\cos\varphi + i\mathrm{Im}(T)\sin\varphi) = C \times T$

Use (1) to focus on fluctuations in imaginary parameters free of normalization concerns.

Use (2) to get complex structure C × (object). Note that exponentiation into phase terms is precisely what occurs in the path integral propagator formalism, and will occur here as well for the emanator formalism, thus the "C ×" complex factor. When drawn upon in the emanator formalism, this method of achieving additional "C ×" complex structure will be forced by the zero-divisor handling (that will give rise to point-like matter with very small phase coupling, thus a highly oscillatory integral, and ties over to foundational aspects of the path integral formalism).

A.2.3.2 Alternate 137th count using Exponential Map

The derivation below follows [2], but with a more succinct accounting of the independent terms.
Consider a general norm=1 bisedenion in list notation: (A,B), where A and B are sedenions. Consider a propagator bisedenion (C,β), $C = (c,\alpha)$, where c is an octonion and α is shorthand for the real octonion $(\alpha,0,0,0,0,0,0,0)$, where α is a real number, and β is shorthand for the real sedenion $(\beta,0,0,0,0,0,0,0,0,0,0,0,0,0,0,0)$, where β is a real number. Using A=(a,b), B=(u,v), and the multiplication rule from Section 2, we have:
$(A,B)(C,\beta) = ([AC-\beta^*B], [BC^*+\beta A])$, where
$AC = (a,b)(c,\alpha) = ([ac-\alpha^*b],[bc^*+\alpha a])$; $BC^*=(u,v)(c^*,-\alpha) = ([uc^*+\alpha^*v],[vc^*-\alpha u])$.
Thus, we have:
$(A,B)(C,\beta) = ([(ac-\alpha^*b\ ,\ bc^*+\alpha a)-\beta^*(u,v)]\ ,\ [(uc^*+\alpha^*v\ ,\ vc-\alpha u)+\beta(a,b)])$,
so,

$(A,B)(C,\beta) = ([ac-\alpha^*b-\beta^*u\ ,\ bc^*+\alpha a-\beta^*v]\ ,\ [uc^*+\alpha^*v+\beta a\ ,\ vc-\alpha u+\beta b])$. Now consider another propagator bisedenion (C',β'), $C' = (c',\alpha')$, and form the product corresponding to the next multiplicative step: $(\ (A,B)(C,\beta)\ )\ (C',\beta') = (\ [(ac)c' - \alpha^*bc' - \beta^*uc' - \alpha'^*(bc^*+\alpha a-\beta^*v)\ ,\ \ldots]\ ,\ [\ldots\ ,\ \ldots]\ )$, where only the first expression at octonionic level ($T=(O_1,O_2,O_3,O_4)$) is shown:

$$O_1 = (ac)c' - \alpha^*bc' - \beta^*uc' - \alpha'^*(bc^*+\alpha a-\beta^*v).$$

At octonionic-level there are 10x9x8/3x2=120 independent terms for 8 octonionic components (labeled a, b, c) plus a separate octonion component (α) and one sedenion component (β), e.g., have 10 choose 3. Also have telescoping terms with repeated real octonion factors, such as with the $a\alpha\alpha'^*$ term (think $a\alpha(\alpha'^*)^n$), which gives an additional 8 independent terms. Also have telescoping terms with alternating real octonion factors and real sedenion factors, such as with the $v\beta^*\alpha'^*$ term (think $v(\beta^*\alpha'^*)^n$), which gives another 8 independent terms. There is one other 'telescoping' term due to repeated octonion right products seen in $(ac)c'$ (now think $((ac)c')c'\ldots..c')$). The change in this term corresponds to an element of the automorphism group on octonions, G2, and as such provides one last independent term, for a total of 137 independent terms at octonion level.

All of the octonion products involve octonions with norms at most unity, and by the normed division algebra rules on octonions, their norm is simply the norm of the individual octonions multiplied together, all of which are bounded by unity, thus their product is bounded by unity. The overall bound for the expression, each individual term being bounded by unity, is therefore simply the counting on the independent terms.

The maximum magnitude of each component of the octonion in the product term is given with a 'channel multiplier' of 137. Also, in seeking the maximum information propagation we require that the real chiral component never cross zero (e.g., stay in its connected $\{\alpha,\beta\}$ quadrant), thus the strictest condition on evaluating evolution might be intuited to be when the imaginary components combine to have real component contribution that is antiphase, e.g., the total imaginary angle is π. The choice of antiphase will used in what follows and will be justified when "C $\times$" allows the antiphase to be understood in the context of the Universal Mandelbrot set [1] position on the negative real axis that gives the maximal magnitude of displacement from the origin: C_∞. We limit the maximum perturbation allowable by the antiphase worst case. At octonionic-level there is thus the channel multiplier: $137 + i\pi$.

A.2.3.3 The {α, π} relation using Exponential Map

The maximum perturbation, referred to as maximum noise in what follows, is first evaluated for a chiral emanation where we take a norm=1 T_{base}=(A,B) and take the right product with T_{chiral} in the form T_{chiral}=(C,β), with product (A,B)(C,β) proven to be unit norm above. In the prior section we saw that there are 137 independent octonion terms at the octonion sub-level of the new unit norm trigintaduonion that results, which leads to 137 independent terms at component level. In order to use the map rules mentioned in the previous section, it is necessary to move from the trigintaduonion, T, space, to the C × T space. This is done in later sections anyway where we consider sums on exp(iT). The exponential function (map) provides a well-defined 'lift' of a hypercomplex (Cayley) algebra from T to C × T. The exponential map also provides a very useful maneuver when working with unit-norm hypercomplex numbers via the generalized deMoivre theorem exp(Im(T))=cos(θ)+Im(T)sin(θ), with the real part recoverable from cos(θ). More details on this follow later but for now, in evaluating the maximum noise allowed we have three structures to adopt: (1) the noise is generalized to be complex (as will be the case for the components themselves once the T→C × T structure is adopted). (2) At component-level, the noise (for maximum noise) is equipartitioned in both real and imaginary parts. (3) Total imaginary noise magnitude is π for maximal antiphase (to be justified later).

(I) Chiral emanation noise: have 137 terms with max unit norm each, for the real part, and for the imaginary part have a "phase angle" β such that 137β=π (here referred to as a phase angle in the sense that the exp(Im(T)) map is being used). The noise magnitude at octonionic-component level is then given by the right triangle with real part = 137 and angle β=π/137, thus maximum chiral emanation noise magnitude is:

$$H = 137/\cos(\pi/137)$$

(II) Achiral emanation noise: now have 29 "free" components, each with 137 independent terms. For maximum achiral emanation we thus have 137 x 29 independent terms that are built from the aforementioned chiral emanation terms (to make achiral). If we equipartition as before, with noise magnitude Hc, we have a "noise triangle" with magnitude (hypotenuse) Hc and with angle $\theta = \pi/137x29$. The imaginary part is then (Hc)sin(θ). As regards the H magnitude separated form (separating out the 'H' factor for now), we have for the imaginary part sin(θ)c. As before, we take maximal noise transmission when all the imaginary parts add to maximal antiphase. Given the equipartitioning assumption, we then simply have the factor 137x29:

$$\sin(\theta)\, c\, (137x29) = \pi \rightarrow c = \theta / \sin\theta.$$

The maximum real noise perturbation that the system can have is then α, where:

$$\alpha^{-1} = \frac{137}{\cos\beta}\cos\theta \frac{\theta}{\sin\theta}, \qquad where\ \beta = \frac{\pi}{137}\ and\ \theta = \frac{\pi}{137x29}$$

$$\alpha^{-1} = 137.03599978669910\,,$$

where the evaluation was done at WolframAlpha to high precision [136] (e.g., higher precision than that reported in earlier work). This matches the experimentally observed value to all 11 decimal places currently known. As of 2002, the measured value of α is:

$$\alpha^{-1} = 137.03599976(50).$$

Note that in quantum field theory the parameters are renormalized at a particular energy scale. Thus choice of energy scale impacts the value of α (as a coupling constant in the classical theory or a perturbation expansion factor in the quantum theory). At 0K we have the extreme low-energy end of the renormalization group (with the largest α value). We are at the 2.7K CMBR, so we have the max α to very high precision. (In studies at high energy scale at LEP, at the energy scale of the Z-boson (91GeV), we get the renormalized value to be: $\alpha^{-1}[M_Z] \cong 127.5$. Note that 91GeV is way above the energy scale of the familiar Hagedorn temperature at ~pion mass=150MeV or 1.7x10^12 K) [444], where hadronic matter 'evaporates' into quark matter.)

A.2.4 Trigintaduonion Emanation: achirality from chirality

There are four chiralities, and for a given chirality (with unit norm) there are 29 dimensions of freedom (10D + 19D of chiraly consistent perturbation). When analytic extension is taken to give maximal information flow, the effective dimension for each of the four chiralities is 29* (detailed in [7]). This clear decomposition into 29* independent effective dimensions is then revealed in the {α,π,C_∞} relation in [7]. The Mandelbrot Set is one of many that encounter the universal constant C_∞. The Mandelbrot set also describes a boundary with 2D fractal dimension at its "edge of chaos" [7]. If driven to similar optimality in approaching a zero-value (a zero-divider issue), we see a two-value zero-crossing specification effectively like a double zero. The parameterization of the zeros of the Emanator at chiral zero-divisor points will thus be as double-zeros.

For what follows we use the simple description of the emanator:

$$T_{chiral}^{(k)} = \begin{cases} ((0,\alpha),\beta) \\ ((\alpha,0),\beta) \\ (\beta,(0,\alpha)) \\ (\beta,(\alpha,0)) \end{cases}.$$

$$\text{Emanation}(\mathbf{T}) = \frac{1}{N} \sum_{k\in\{4x72\}^n} \mathbf{T} \bullet \boldsymbol{T}_{chiral}^{(k)} = \frac{1}{N} \sum_{K\in 4\ chiralities} \mathbf{T} \bullet \overline{\boldsymbol{T}}_{chiral}^{(K)}$$

If working with non-split T's, then we restrict to emanations that are perturbations of unity:

$$T_{chiral}^{(k)} = \begin{cases} ((0,\alpha),\beta) \\ ((\alpha,0),\beta) \\ (\beta,(0,\alpha)) \\ (\beta,(\alpha,0)) \end{cases}, where\ T_{chiral}^{(k)} = \mathbf{1} + i\boldsymbol{\delta}.$$

From unit norm we have $\alpha^2 = 1 - 0^2 - \beta^2$, with ± sign choice on α, similarly for β.

If working with split T's (bi-sedenions, etc.), then we have manifest Lorentz Invariance. So, it is often convenient to work with split-T' since this is manifest from the outset.

Issues with zero-divisors

Suppose we add the rule that emanation may not proceed when a particular chirality is zeroed-out, in other words:

$$\mathbf{T} \bullet \overline{\boldsymbol{T}}_{chiral}^{(K)} \neq \mathbf{0}.$$

For 'normal' numbers this goes without saying, since for real numbers if we have $r_1 \times r_2 = r_3$ then $r_3 \neq 0\ if\ neither\ r_1 = 0\ or\ r_2 = 0$. This holds true for the Real, Complex, Quaternion, and Octonion numbers. This does not hold true for Sedenions or higher. For sedenions the dimensionality of the zero-divisor event is mostly constrained, while for trigintaduonions it is significant. If such zeros were eliminated from the emanator description by using analytic extension component-wise (on

29* effective components) we see how a description devoid of matter (pure static field with no source or sink) might acquire matter by way of extending to a maximal domain of analyticity be removing zero-divisor events (a Wick transformation from real dimensionless action to pure imaginary action that is dimensionless but consisting of a dimensionful ratio). For further discussion along these lines, see Book 7 [1].

A.2.4.1 Achiral T-emanation has 29* effective dimensions

Let's estimate of the effective dimension of information transmission in an achiral T-emanation process. There are 4 chiralities, so to get an achiral emanator candidate, minimally need a "4-card deck" to emanate in the four chiralities, with emanator equal to normalized sum. The actual deck appears to require a normalized sum over sub-chiralities, as will be explicitly enumerated in what follows. Here are the four chiralities with real fluctuation noise shown:

$$\begin{gathered}(\,(\,(\mathrm{O}[0]\pm\delta,\ldots),\alpha\pm\delta),\beta\pm\delta)\\(\,(\,\alpha\pm\delta,(\mathrm{O}[0]\pm\delta,\ldots)),\beta\pm\delta)\\(\,\beta\pm\delta,(\,(\mathrm{O}[0]\pm\delta,\ldots),\alpha\pm\delta)\,)\\(\,\beta\pm\delta,(\,\alpha\pm\delta,(\mathrm{O}[0]\pm\delta,\ldots)\,)\,)\end{gathered}$$

where α is a real octonion and β is a real sedenion, and Tem is an equal weight sum of the action of each of the sub-chiral propagations on the base T, with the fluctuations indicated each done separately. We have the constraints $\alpha \neq 0$, $\beta \neq 0$, and common octonion O not pure real.

Each of the δ's is an independent fluctuation corresponding to its own sub-chiral emanation, but no subscripting on δ's is used or shown. There are thus 9x2x4=72 independent *imaginary* noise fluctuations to consider in the exp(Im(Tem)) evaluation (that automatically provides unit-norm). The real noise fluctuations in the real (first) component are, thus, not counted. If our definition for Tem entails only one card being dealt, then the sum over those possibilities is the sum

$$\mathbf{T}\bullet\mathrm{T_{em}} \equiv \mathrm{Emanation}(\mathbf{T}) = \frac{1}{72}\sum_{k\in\{72\}}\mathbf{T}\bullet\boldsymbol{T}^{(k)}_{chiral}$$

For one-card, or a one-step, emanation, with real components and real noise, this makes sense from the counting shown, and it's what we use going forward. Using this will allow an entirely separate method for evaluating α (here at the one-card hand approximation). This will be done by determining the effective dimension 29*>29 of maximal information propagation (or maximal noise fluctuation). Before moving on, however, let's examine what happens when we allow complex noise fluctuations as this will trivially be allowed when we consider C × T via exp(iT) in later discussion anyway.

A.2.4.2 Effective Deck size is 72, which is consistent with the {α,π,C_∞} relation

Maximum information transmission involves a complex extension to the T components and their noise fluctuations, but in doing this it must retain emanation structures such as the octonionic triple that occurs in previous expressions (starting with the proof of the T_{chiral} solution itself), which leads to the counting that gives 137 independent terms, etc. Thus, the maximal complex extension on the noise is that it remain real in the octonion components:

$$(((\mathrm{O}[0] \pm \delta, \dots), \alpha \pm i\delta), \beta \pm i\delta)$$
$$((\alpha \pm i\delta, (\mathrm{O}[0] \pm \delta, \dots)), \beta \pm i\delta)$$
$$(\beta \pm i\delta, ((\mathrm{O}[0] \pm \delta, \dots), \alpha \pm i\delta))$$
$$(\beta \pm i\delta, (\alpha \pm i\delta, (\mathrm{O}[0] \pm \delta, \dots)))$$

The first chiral T component is where new imaginary terms might arise (the others are already counted since in imaginary components). We see there are six more, so the deck is now78. In application, as we will see, those added six are precluded due to constraints such that the effective deck size (impacting the sums and numbers of independent terms in the emanator definition) remains 72.

All noise terms will be treated additively, including terms in different imaginary components as well as imaginary noise terms in the real component. The criterion for max noise (in-phase constructive interference) gives the extreme of linear additivity. (Not like Gaussian statistical noise that adds in quadrature.) Also note that the discussion in terms of "noise transmission" and "information transmission" will be used almost interchangeably, whenever one description or the other best suits the analysis it will be used. Note that with this kind of noise analysis we can effectively shift around T noise terms associatively. Also note that application of the Kato-Rellich theorem [445,446] is related to the noise budget analysis done here focusing on first order terms.

There are 137 independent tri-octonionic terms in each of 29 free components indicated by a particular chirality (within the 32 components of a general trigintaduonion). This is a nontrivial result since (T_{chiral} • T_{chiral}) is no longer T_{chiral} type (but still T_{norm1} type), so direct expansions are needed to identify the number of independent terms and this is briefly described below, with more detail in [1].

Obtaining an achiral emanation from a collection of chiral emanations requires that all chiralities be summed over (there are four) as well as sub-chiralities (there are 72). Noise analysis requires collecting of first-order terms. Analysis of noise transmission indicates 29* dimensions, where:

$$29^* \cong 29 + \left(\frac{4\pi}{72}\right)\left[1 + \left(\frac{\pi}{137 \cdot 29}\right)\left(\left(\frac{\pi}{72}\right) + \left(\frac{3}{72}\right)\right)\right]$$

The above result was obtained in [132] to describe the 72-card chiral 'deck' of chiral emanation products for a "single-step" emanation. In the new Results to follow this is reviewed and elaborated further.

A.2.4.3 'Edge of chaos' maximal perturbation hypothesis [7]
Consider the 'edge of chaos' maximal perturbation in each of the 29* dimensions to be at position C_∞, which is on the negative real axis, i.e., at π rotation to have -1 factor, ***thus at maximal antiphase***. This results in the relation for maximal perturbation at maximal antiphase (maximum reference angle with sign chosen positive by convention) has a lower bound on α given by:

$$\alpha^{-1} = \left(\sqrt{C_\infty}\right)^{29^*}.$$

where

$$C_\infty = 1.4011551890920506004 \ldots$$

This ties $1/\alpha$ to the second Feigenbaum constant C_∞ in the context of the Mandelbrot set. It is well known that the Feigenbaum constants are universal, and part of a description of a universal transition to chaos regime. The Mandelbrot set is also universal [42], and maximal in that its fractal boundary has maximal fractal dimension of 2 [42], a detail that will be important in the meromorphic matter description given later.

For C_∞, most references only provide $C_\infty = 1.401155189 \ldots$, and a higher precision tabulation is not readily found, so use is made of the relation

$$C_n = a_n(a_n - 2)/4,$$

together with the tabulation on a_∞ [447]:

$$a_\infty = 3.569945671870944901 8 \ldots$$

The resulting C_∞ is:

$$C_\infty = 1.401155189092050600 4 \ldots$$

The resulting α^{-1} is:

$$\alpha^{-1} = 137.035999333701982 63 \ldots$$

A.3 Implication of MIE

A.3.1 MIE requires a complex Hilbert Space

As mentioned previously, according to [45], a complex Hilbert space is selected by the quantum deFinetti theorem, since it is required for information propagation (and thereby consistent with the maximum information propagation concept in its selection). Because it's a complex Hilbert space, this explains why the path integral operates in a complex space, even though the underlying universal algebraic construct from which it is emergent is hypercomplex to the level of the trigintaduonions.

From [45], a simple derivation shows why the quantum deFinetti Theorem requires amplitudes to be complex. Suppose f(n) is the number of real parameters to specify an n-dimensional mixed state. For real amplitudes f(n)=n(n+1)/2, for complex amplitudes f(n)=n^2, and for quaternionic f(n) = n(2n-1). For propagation, etc., need f(n1n2)=f(n1)f(n2), which only works for complex amplitudes.

A.3.2 Time is analytic and Matter is meromorphic

A variety of efforts have been made to find a definition of time that is somehow implicit to the main QFT and GR formalisms, whether it be a choice of vacuum for QFT in curved spacetime (and even if the spacetime is not curved [448]) which is indirectly a choice of time. Or seeking an internal time-reference in a full-GR quantum minisuperspace analysis of dust-shell collapse [449]. Or in seeking a notion of time in full general relativistic (GR) models, in the equilibrium sense, with an assumption of euclideanizability [450,451]. For the latter, the self-consistent stable solutions that were indicated showed the general utility of the euclideanizability hypothesis on emanation/propagation solutions in general (that is especially relevant, or interpretable, when the system is in equilibrium). In none of these efforts, however, was there success in identifying some internal notion of time, time, it seems, is an added construction, and this is consistent with the results shown here, where we find that time is likely analytic and an emergent construct.

In Book 7 [1], we see matter as meromorphic residue precipitation, in amounts of one quantum given by a precursor to Planck's constant h*. The meromorphic residue winding number is also notable in that it gives an integer that stays constant in the meromorphic region. This raises the possibility that elementary particle attributes might encode by way of different winding numbers, with reference to their different winding numbers at residues, but this will not be discussed further here.

A.3.3 Emergent Evolution and Emergent Universal Learning

We see that the definition of the emanator process is not known, but that consistency arguments (such as achirality constructed from a collection of chiral emanations) lead to a certain set of forms. And that consistency with the $\{\alpha, \pi\ C_{\infty}\}$ relation imposes further constraints on the form of the emanator. What is hypothesized is that the emanator is selected for maximal information transmission, thus emergent itself under that criterion. Let's now consider the maximal information transmission idea from the receiving end, e.g., maximal information receiving, or learning, in this context. If we turn to the information geometry analysis of learning in neural nets [191-193,452,453] (which uses differential geometry) we obtain a fundamental origin for statistical entropy (Shannon Entropy), and we identify optimal learning processes, based on expectation/maximization, that involves two-steps (as the name suggests) that may be done according to two fundamentally different conventions, e.g. the optimal learning involves four types of step, or is doubly chiral, consistent with the emanator 4-chiral processes described [1]. The potential applications of these results to Trigintaduonion encoded neuromanifolds is beyond the scope of this paper.

A.3.4 Objective Reduction, Zero-Divisors, and possible origins of Planck 's constant

A new mechanism for objective reduction [454,455] is also indicated by the way π enters the theory as a maximum anti-phase amount comprising part of the maximal perturbation propagation. Consider in the context where there is a 'classical' trigintaduonion path in a congruence of paths (a flow-line description). On the classical path in the congruences, we have α calculated using a $+\pi$ maximal anti-phase, but this could also occur with $-\pi$ maximal anti-phase as well, thus we could have a $\pm\pi$ phase toggle when a zero divisor is encountered in the 32D propagation (given the perturbations extending outside the 10D somewhat into the entire 32D). The zero-divisor discontinuity requires the field to reformulate a new 'consistency' with the 32D algebraic propagation (and 64D and higher, as well), which could have the result that since the prior π phase had the discontinuity, then it must toggle to the other, negative, phase, e.g., objective reduction may occur as a zero-divisor phase-toggle event.

A.3.5 Where's the geometry?

The geometry side of emanation theory does not result from the action of the repeated emanator product directly, but from the accumulated product in the T base that results. Geometry is, in effect, emergent (projected) on the T_{base} 'space' of the T_{em} product action. Geometry appears as a manifold construct in both space-time curvature (where it is locally given with the standard model action) and as an intrinsic entropic property, via neuromanifold 'geodesic' motion being equivalent to, and possibly the origin of, the minimization of the relative entropy (and maximization of entropy, the 2nd Law). Setting aside thermodynamic issues in this discussion, this puts the Lagrangian formulation with standard model terms, and Hilbert action for GR, into better perspective. The representation of the geometry via the Hilbert action for GR suffices with maximal extension in whatever causally connected domain of interest. So, we've got the existing QFT in CST space-time formulations in the black hole exterior, for example. We may have the resolution at the black hole horizon causal boundary via String Theory on the surface (using Ads/CFT relation and related holographic hypothesis [456,457]).

We describe repeated chiral product action on the trigintaduonion spinor space. The emanation process, consisting of a chain of chiral trigintaduonion products, leads to a Lagrangian variational formalism *with the standard model*. The origins of the parameters of the model are beginning to be understood as well. Apparently state information memory/inertia is carried via the manifold curvature response to the matter density, where 'G' is the linkage for the balance on this 'learning' process. Presumably the G learning rate is set for optimal learning, e.g., maximal information flow, and as such its value may eventually be clarified theoretically.

A.4 Results

A.4.1 Two-card Hand consistent with 137 and Neuromanifold Two-step Learning

MIE may guide our choice of 'hand' size dealt. Consider that one chiral emanation step (multiplication) off a base trigintaduonion will only have

78 independent tri-octonionic noise terms introduced. With a second step we 'flood the channels' and get the full assortment of 137 tri-octonionic noise terms (as shown in the derivation that obtained the 137 terms in [7]). Thus, a minimum of a two card hand must be considered. Having achieved a maximal perturbation scenario, what need for further cards in the hand, i.e., or further multiplicative factors in the pre-achiral normalization step?

Previously, when speaking of "Single-step achiral" we presumed initial conditions for the noise that effectively performed an average two-card 'hand' emanation analysis. If not in such an approximate form the ***{α, π}*** relation is recovered. The approximate form, however, expresses a new hypothesized relationship: $\alpha^{-1} = \left(\sqrt{C_\infty}\right)^{29^*}$, that is itself verified to many decimal places.

Two-card-hand and two-step emanation jibes with two-step neuromanifold 'learning via 'em' and indicating relative entropy as optimal in this process (similar to choice of Euclidean distance in Riemann spatial change, here we have relative entropy as local entropic measure (with fixed reference, this then gives Boltzmann's entropy.

A.4.2 The Chiral-Extension Cayley-Family relation

Recall that the unit-norm propagation was extended from octonion-level to trigintaduonion-level by introduction of chiral trigintaduonion emanators. The mathematical construct at the trigintaduonion-level that results can be represented in terms of tri-octonion products, thus all of the octonion properties are thereby inherited with such a construct since it is represented in terms of octonions, most notably a norm (here unit norm is of interest physically, for repetitive 'propagation'). The same double-chiral extension can be done at any Cayley level, and we thereby have a Cayley-Family extension property, via use of chiral Cayley emanation only, that is general.

Associativity Extension most likely Candidate for Renormalization and Thermality reification

Application of the Chiral-Extension Cayley-Family relation is first considered for associativity. In the Cayley Algebras, the highest algebra that retains the property of associativity is the quaternions. If we consider extending this associativity to higher -order algebras in the context of chiral propagation we can do this up to algebras at the level of the octonions and the sedenions. Consider that ***unit-norm*** trigintaduonion emanation will reduce to sedenion propagation in a general sense, and it is in this space that we can now add another sedenion element that is associative.

A.4.2.1 Sedenion Associative Element and the Axiom of (Renormalization) Choice

The description of the renormalization process involves counter-terms. The emanation/selection of the standard gauge field group, and other elements (described in Section X), seems consistent with the Standard Model (and certain extended versions), but makes no mention of counterterms. Rather than add counter-terms as an external regularization element to the theory, here we can consider them as a more complete manifestation of the existing Emanator theory (again the MIE hypothesis justifies this). So, let's suppose there really is chiral trigintaduonion emanation, we can now see where a side-version of the theory presents that is sufficient to provide a (reified) version of the renormalization terms needed to renormalize the theory. Note that the application of the renormalization 'cancellation' is a fundamental application of the Axiom of Choice. More accurately, the universe that is hypothesized to be MIE, implicitly selects for the existence of the Axiom of Choice (and well-ordering foundation of mathematics) in its choice of renormalization cancellation.

A.4.2.2 Octonion Associative Element and Complex Periodic Time

If extension of associativity on a sub-class of chiral-trigintaduonion emanations was possible, via the aforementioned effective sedenion associativity on chiral propagation, what of the indicated octonion-associativity extension? At the level of the octonion we can see a more direct manifestation of Lorentz invariance, including terms that are spatial or temporal (see generalized Lorentz invariance representations in Book 4 [82]), so suppose it is here that the theory creates ties between the real time quantum dynamics on a manifold and the complex-time statistical dynamics on a neuromanifold. These ties are shown to exist during equilibrium analyses in a number of applications (ss Book 6 [34]), and in thermal quantum field theory applications (see Book 5 [35]), but here we speak to this being a more fundamental attribute, revealed in the indicated circumstances, yes, but also present in general (albeit perhaps not in a useful form).In essence, we use our second wish/demand (we only get two) of the associative-extension genie that it explain the recurrence of the appearance of a fundamental thermality by way of time having a complex periodicity proportional to inverse system temperature.

‘Contact’ Associativity from Higher-order Cayley Algebras (64 and 128 orders for example) was suggested previously, involving use of limit processes and infinitesimal processes on the non-propagatable orders (even chiraly) above 32-order trigintaduonion, to be at zero displacement from the 32-algebra (thus ‘contact’), but still imparting possible nontrivial contributions, such that the above renormalizability and complex-time relations might be obtained. Here we find a better route to this result via the chiral-extension property.

A.4.2.3 Octonion Commutative Element and Analytic Time

If we are considering operations reduced to octonion-level, there are also chiral extensions on the commutative complex numbers to arrive as a subset of commutative octonion-level operations in the unit-norm chiral emanation process. This allows further modification to the octonions as regards the internal time parameter analyticity, not just its imaginary-time periodicity. Together these chiral extensions may provide for the analytic time that is observed in the many successful applications of such a notion (required in path integral descriptions of bound states, for example).

The multiplication properties of the Cayley algebras are: Commutativity, Associativity, Alternativity, and Power Associativity. As you go to higher order algebras you lose properties. The Complex numbers are the last to have commutativity, the quaternions the last to have associativity, and the octonions the last to have alternativity. All the algebras have power associativity. Thus, the chiral extensions described are now exhausted since the interesting cases to extend, from complex, quaternionic, and octonionic, have now been considered.

A.4.3 The Extended Standard Model predicted by Emanator Theory

According to the Standard model of Particle Physics, the Universe consists of point-like spin-1/2 fermions that reside in two families: the Leptons and the Quarks. These particles exist in a geometry with a gauge field, where they interact through that gauge field according to local gauge invariance and indirectly through the geometry according to general relativity. Even more indirectly, the gravitational interaction via mass has mass itself determined through exchange of spin-0 Higgs particles.

The local gauge invariance of the Standard model is:

$$U(1)_Y \otimes SU(2)_L \otimes SU(3)_C.$$

Let's consider each part separately:

$U(1)_Y$→ gives rise to the electromagnetic force between charged particles.

$SU(2)_L$→ gives rise to the force between left-handed particles.

$SU(3)_C$→ gives rise to the strong force that operates on particles (hadrons) that carry color charge (three types).

In Book 5 [35] we see that each gauge group listed above derives from the general local gauge invariant group form:

$$SU(N)_L \otimes SU(N)_R.$$

For $N = 1$, we have the trivial case (no left or right): $U(1)$.
For $N = 2$, we have the asymmetric case of only left: $SU(2)_L$.
For $N = 3$, we have the symmetric case of left-right, or simply a fixed ratio: $SU(3)$.

In Book 5 we see that there is mixing between the $U(1)$ and $SU(2)$ parts to give the actual electroweak theory observed:

$$U(1) \otimes SU(2) \rightarrow \; U(1)_Y \otimes SU(2)_L.$$

The mixing introduces universal (globally gauge invariant) 'mixing angles' that number among the fundamental parameters of the theory (along with particle masses and a few other constants).

In Book 7 [1] we see that Emanator Theory predicts the form of the local gauge invariance as a general form, to be exactly that observed: $U(1)_Y \otimes SU(2)_L \otimes SU(3)_C$. Emanator theory also predicts a theory governed by 22 parameters, not 19 as currently listed for the Standard Model.

As strange as the form $U(1)_Y \otimes SU(2)_L \otimes SU(3)_C$ might seem, it is so far only applied to the interaction element of te theory. Let's now consider an interaction with what? The answer resides in the representations of the indicated groups (the left-handed particles interacting according to $SU(2)_L$ will be seen as interacting visa the weak force, etc.). So, saying there is a local gauge invariance is to say that there will be particles according to the irreducible, independent, representations of these groups. Thus, for any chosen representation there will be a collection of particles predicted. This is precisely what is observed experimentally. Thus, we not only know the groups $SU(2)_L$ and $SU(3)_C$, say, we also know their specific representations to obtain particle numbers and groupings observed experimentally.

So, the Standard Model is actually indicting the Model pair of Local Gauge Field and Representation $\mathcal{R}$:

$$\{U(1)_Y \otimes SU(2)_L \otimes SU(3)_C;\ \mathcal{R}\},$$

and it (the Standard Model) then says "times three", where the representation has there generations or copies:

$$\{U(1)_Y \otimes SU(2)_L \otimes SU(3)_C;\ 3 \times \mathcal{R}\}.$$

For the fundamental Leptons and Quarks this is shown in tabular form with the symbols for the particles as:

	electroweak left-handed Leptons	electro-only right-handed Leptons	Electro-only Quarks
1st Generation	$[e^-{}_L, \nu_{e,L}(m=0)]$	$[e^-{}_R, ---]$	d, u
2nd Generation	$[\mu^-{}_L, \nu_{\mu,L}(m=0)]$	$[\mu^-{}_R, ---]$	s, c
3rd Generation	$[\tau^-{}_L, \nu_{\tau,L}(m=0)]$	$[\tau^-{}_R, ---]$	b, t

Note that there are no right-handed neutrinos in the Standard Model and the left-handed neutrinos are massless. We already have ample evidence that the left-handed neutrinos are massive, so we know the Standard model will extend in this regard, which leaves the supposedly non-existent right-handed neutrinos. There probably are right-handed neutrino but if "electro-only" and having no charge, what results is a particle completely decoupled from the other matter, except gravitationally, precisely what describes dark matter. More on this will follow once we go from the 19 parameter model to the hypothesized 22 parameter model, as this suggests limits on extensions to the theory, as does cosmological data, which suggests a possible 4th neutrino and no more scenario consistent with the cosmological evolution, unless the other neutrinos are very massive (all probably the case).

With the Standard Model quantum field theory structure, with scatter results according to the indicated local gauge field and particle representation, we find additional conservation rules. One of these rules describes the conservation of Lepton number at each generation, in other words:

$$L_e = N(e^-) + N(\nu_e) - N(e^+) + N(\bar{\nu}_e)$$

is a constant, and the same for constants for the muon and tau generations. There is also conservation of baryon number and the familiar conservation of charge.

The Standard Model is a renormalizable/renormalized quantum field theory, and while this poses no difficulty with the $SU(2)$ elements alone or $SU(3)$ alone, here we have a product gauge and, what's more, it has handedness with use of $SU(2)_L$ not $SU(2)$. Not surprisingly, anomalous terms arise in he renormalization effort and for the renormalization to work, additional constraints are imposed on the structure of the theory. Most notable is that at a generational level across both leptons and baryons there should be a charge sum of zero. This is observed and was used to predict missing quarks (charm) in the early discovery process.

The structure identified by the standard model and the high decimal number agreement on key results from quantum field theory together provide a theory that can't be displaced directly by anything better, and only minimally augmented (with addition of right-handed neutrino), so trying to find a unified theory is impossible from 'within' the theory, and Godel's incompleteness theorem also suggests the hopelessness of such a task. This, then leaves no choice but to look for an encompassing theory that projects the quantum theory with the Standard Model structure in its entirety. And this is what is attempted with emanator theory (Book 7 [1]).

The Standard Model can't explain the three generations of matter, the origin of the local gauge group with its odd product form, the dimensionless constant alpha, and certainly can't explain the extension to massive(light) neutrinos, or the possible extension to massive (dark) right-handed neutrinos. In Emanator Theory (Book 7 [1]) we have:
(1) The local gauge $U(1) \otimes SU(2) \otimes SU(3)$ is projected by the theory. Why it should have three generations and why it should by asymmetric with use of $SU(2)_L$ is due to those choices providing the maximal packing of the Light matter sector (for maximal complexity information flow, an aspect of the MIE hypothesis), within he constraint of a 22-parameter emanation process.
(2) The 22 parameter, constants of the motion (emanation), result helps to constrain the particle representations such as to stay at 22-parameters, yet have maximally complex interaction.
(3) The MIE Hypothesis indicates the given generational structure given the constraint of working with the indicated product algebra. It also suggests any 'fine-tuning' on gravitational constant G might be dominated by the dark matter sector and its contribution to the evolution of the universe. Such fine-tuning would be most powerful if it indicated a fractal scale invariance property (see fractal G discussion in [10]).

Before considering the 22-parameter Emanator theory prediction of an extended Standard Model, let's first recount the 19-parameter Standard Model, which I'll separate into four groups:

> (I) 9 Yukawa coupling constants (masses) for the charged fermions
> (II) 5 constants for Weinberg Angle and the CKM matrix (with three mixing angles and CP-violating phase)
> (III) 3 Constants for electromagnetic coupling (alpha), for strong interaction (g3), and strong CP-violating phase ($\theta_3 \approx 0$).
> (IV) 2 Higgs parameters: Mass and Vacuum Expectation

If we allow for the left-handed neutrinos to have mass, then we get 3 more masses and another 4 constants for the PMNS matrix (three mixing angles and a CP-violating phase):

> (V) Extended model: 7 more constants → We, thus, have 26 parameters.

Let's update out table with this extended version of the theory:

	electroweak left-handed Leptons	electro-only right-handed Leptons	Electro-only Quarks
1st Generation	$[e^-{}_L, \nu_{e,L}(m \neq 0)]$	$[e^-{}_R, ---]$	d, u
2nd Generation	$[\mu^-{}_L, \nu_{\mu,L}(m \neq 0)]$	$[\mu^-{}_R, ---]$	s, c
3rd Generation	$[\tau^-{}_L, \nu_{\tau,L}(m \neq 0)]$	$[\tau^-{}_R, ---]$	b, t

There is now a problem. In order to maintain renormalizability, if there are left-handed neutrinos with mass there must be right handed neutrinos with mass [458]. So, just how sure are we about neutrinos having mass? There is strong evidence not only for neutrino mass, but for neutrino family dynamics (here seen as oscillation between neutrino mass states). We'll see more evidence of generational family kinematics/dynamics in a later section. Neutrino family dynamics was first seen in measurements indicating that neutrinos spontaneously changed flavor (the Solar neutrino experiments [459]), which not only indicates they have mass, but allows us to determine the mass differences.

Clearly the Extended Standard Model needs to be extended further, to allow for the massive right-handed neutrino that is hypothesized to have no weak interaction like its right-handed electron cousin (and no electric interaction since no electric charge and no strong interaction since no color charge, thus 'dark' matter). Such a right-handed neutrinos (with no charge) can act as their own antiparticle (which is consistent with formation of "Bright" supermassive Black Holes in the early universe as seen with Webb). Furthermore, in addition to the Dirac mass relation to the left-handed neutrino it also has a Majorana mass term not tied to the Higgs mechanism [460] What results is that instead of mass ${e^-}_L$ equal to mass ${e^-}_R$ here we have

$$m_{\nu_{e,L}} \propto \frac{1}{m_{\nu_{e,R}}},$$

which is known as the see-saw mechanism [461] Thus the very low-mass left-handed neutrinos indicate very large mass right-handed neutrinos. Large mass neutrinos are an excellent candidate for cold dark matter, precisely what is needed to complete the Cosmological Standard Model.

Given the renormalization constraints and the observation of neutrino mass we arrive at the following updated Model:

	electroweak left-handed Leptons	electro-only right-handed Leptons	Electro-only Quarks
1st Generation	$[{e^-}_L, \nu_{e,L}(m \ll m_e)]$	$[{e^-}_R, \nu_{e,R}(m \gg m_e)]$	d, u
2nd Generation	$[{\mu^-}_L, \nu_{\mu,L}(m \ll m_e)]$	$[{\mu^-}_R, \nu_{\mu,R}(m \gg m_e)]$	s, c
3rd Generation	$[{\tau^-}_L, \nu_{\tau,L}(m \ll m_e)]$	$[{\tau^-}_R, \nu_{\tau,R}(m \gg m_e)]$	b, t

which is described by the 26 parameter theory indicated for massive left-handed neutrinos (if we assume that the right handed neutrino masses can be determined from the left-handed neutrino masses).

The standard counting to arrive at the 19 parameters (or, now, 26) includes $\theta_{QCD} \cong 0$ as a parameter of theory and its nearness to exactly zero is often referred to as the Strong CP problem. Going forward this is removed as a concern. The 'fine-tuning' that selects $\theta_{QCD} = 0$ is considered to be the same as that which selects the generational number, to arrive at 22 parameters while respecting the local gauge field constraint. Thus t is part of the MIE optimal selection.

Also included in the 19 parameter count is the U(1) gauge coupling g_1 and the SU(2) gauge coupling g_1, and these coupling are related to alpha by:

$$\alpha = \frac{1}{4\pi} \frac{{g_1}^2 {g_2}^2}{{g_1}^2 + {g_2}^2}$$

Now, the emanation's projection of the quantizable Lagrangian theory involves various constants (22) in the resulting quantization. In that process, however, alpha already exists, in fact, from the derivation of alpha, we see that it establishes maximal perturbation starting at the level of the emanation process itself. Thus, alpha doesn't number among the 22, and the relation between coupling constants involving alpha means only one of those coupling constants is independently specifiable, so only one should be counted. At tis juncture we've reduced the count on 26 parameters to 24.

Notice in the prior counting, to arrive at the count of 26, we extended the count to be the same with dark neutrinos included since the dark neutrino mass was assumed to have some fixed (inverse) relation to the light neutrino mass. Let's not make this assumption, but retain the same mixing angle matrix, to now have 3 more masses in the theory (adding to the 24 count we now have 27).

In order to reduce the 27 parameter count to 22 we need to remove an 'overcount' of 5. Such an overcount would occur if there was a family-relationship generationally, for all masses. Consider such a mass relationship for the electron-muon-tau family, it is already known to exist and is known as the Koide relation. The Koide relation [462] was first observed for the three massive leptons currently known:

$$\frac{m_e + m_\mu + m_\tau}{\left(\sqrt{m_e} + \sqrt{m_\mu} + \sqrt{m_\tau}\right)^2} = \frac{2}{3}$$

To a lesser extent this relation is satisfied for the quarks as well, particularly for the three most massive, where the value is 0.6695. The problem with a simple application to the quark masses is that they are dependent on energy scale. A theoretical explanation for the Koide relation describes how this relation might exist for the masses of a given generation (or family group) [463]. As mentioned with neutrino oscillations, their oscillations involve the existence of such a family relationship in the quantum field theory implementation, so the hypothesis that it exists for each column in the table is not such a stretch. There are five columns of (independent) masses (the right handed electrons not independent of the left handed electrons removes them from the count). Thus, the five groups of three represent five groups with only two independent mass parameters, and the overall parameter count is thereby reduced by 5 from 27 to 22, as needed to be in agreement with Emanator theory. Thus Emanator theory, with the MIE hypothesis, would select for family group dynamics as observed with the neutrinos since it is what allows the maximal particle set within the 22-parameter constraint.

A.5 Conclusion

Maximal information propagation as an emergent construct appears to require two forms of propagation, an early hypercomplex 'emanation' that involves a chiral 10D propagation in a 32D trigintaduonion space, and standard propagation with complex propagators (consistent with the quantum deFinetti relation) operating inside the geometry and gauge field that is projected. From the 'emanation' stage we see the maximum dimensionality and fractal limits provide the fundamental constants that then imprint upon the emergent geometry and gauge field, including giving rise to the constants α. The origin of α has been a long-standing mystery. So much so that the central role of α in modern physics is literally engraved in stone, the tombstones of Sommerfeld (which displays $e^2/\hbar c$, which is α) and Schwinger (which displays $\alpha/2\pi$) for example. Its origin has eluded physics for over a century, and appears to reside in the algebra of trigintaduonions.

Emanator Theory results from a hypothesized maximal information propagation and this means maximal analyticity, maximal domain, etc. As a process, Emanator theory is also hypothesized to operate up to "the edge of chaos" to permit maximal perturbation (noise) domain. When taken with the results showing that Emanator theory is Martingale, thus has well-defined limits, we then must wonder if there are well-defined multi-scale (fractal) limits. In other words, is there a relation that would tie the micro scale constant to fundamental constants (as they are counted in the 22) with the cosmological scale 'constants' that have settled out, at macro scale, in the current evolution of the Universe? In this context, the Gravitational constant G is hypothesized to be a multiscale fractal coupling parameter.

In seeking a deeper theory we build on the sum-on-paths with propagator formulation to arrive at a sum-on-emanations with emanator formulation. Propagation in a complex Hilbert space, however, in a standard QM/QFT formulation, requires the propagator function to be a complex number (not real or quaternionic, etc., [45]). This prohibits what would otherwise be an obvious generalization to hypercomplex algebras. In order to achieve this generalization, we have to introduce a new layer to the theory, one with universal emanation involving hypercomplex algebras (trigintaduonions) that is hypothesized to project to the familiar complex

Hilbert space propagation with associated fixed elements (e.g., the emanator formalism projects out the observed constants and group structure of the standard model). The 'projection' is an induced mathematical construct, like having SU(3) on products of octonions, but here it we be the standard model U(1)xSU(2)xSU(3) on products of emanator trigintaduonions. Thus, a unified variational formulation is posed, one that arrives at alpha as a natural structural element, among other things, uniquely specified by the condition of maximal information emanation.

A 'deeper' phase of universal evolution is described by a theory of emanations, where mathematically invariant emergent structures appear:

$$\boldsymbol{emanation \rightarrow propagation \rightarrow trajectory}$$

At the emanation to propagation emergence, one of the emergent constructs is the familiar path integral based on standard (unitary) propagators in a complex Hilbert space.

We have α, 10,22,78,137 as parameters resulting from analysis on a single path maximal information flow construct, where the number 22 corresponds to the number of emergent parameters in the description of the propagating construct (exact derivation of the 22-parameter in [1]). In addition, the time choice is emergent via a multi-path construct, along with the propagator construct, and is coupled in both time step and imaginary time increment. The formulation is inherently embedded in a higher dimensional complex space, thus all of the QFT complex analysis analyticity methods are valid as the assumptions made are now part of the maximal information flow emergent construct.

Appendix B. Evolution, Compartmentalization, and Nucleic Acid Genomes

This section derives from a larger biology analysis presented in [127].

B.1 Emergence of cell/virus and standard dogma: DNA→mRNA→protein

The classic central dogma of biology describes how information encoding in the form of a DNA polymer (or collection of DNA polymers, e.g., chromosomes) transcribes to messenger RNA (mRNA) polymers, which are then translated to protein using a three-base encoding scheme (see Fig. B.1). The three-base encoding scheme leads to the discovery of anomalously long ORF encodings, which is the basis of the ORF-finder algorithm described in [127].

For an organism's genomic DNA information repository the transcription from DNA to mRNA can be done in multiple ways from the *same* section of DNA, e.g., different overlapping reads are possible where different three-base codon framings step across the same DNA encoding region (see Fig. B.2). For some prokaryotic organisms this overlap encoding can be quite significant [127] (see Ch. 3 for details, and this overlap encoding by frame shift is effectively doubled for duplex DNA genomes (with encoding in the complement strand), as well as duplex RNA genomes, and ssDNA and ssRNA genomes that have an antisense (reverse complement) encoding read-out by way of an intermediate duplex form. So, for prokaryotes a high degree of overlap encoding, for both forward and reverse reads, is already prevalent at the level of the genome, whether the genome is single stranded or duplex, or DNA-based or RNA-based.

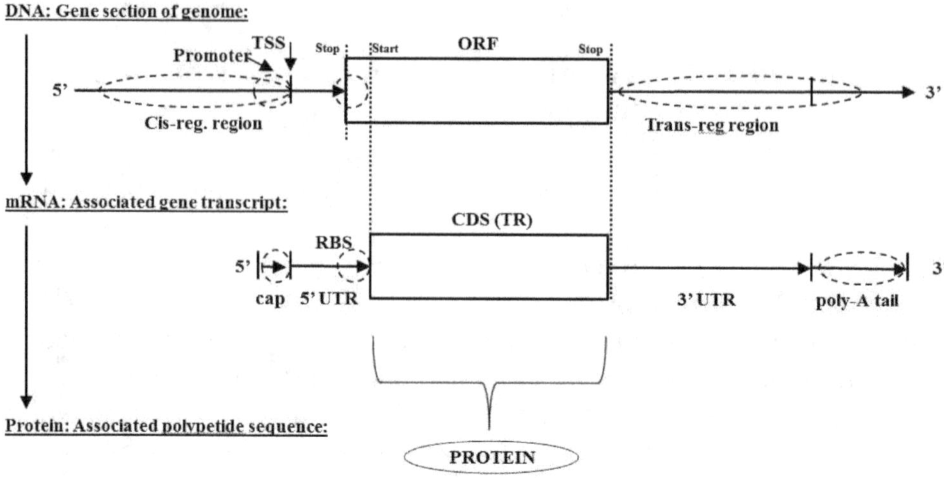

Figure B.1. Central Dogma: DNA → mRNA → Protein (monicistronic).

In Ch. 3 we see that the *C. trachomatis* genome is half coded on the forward strand and half on the reverse, with very little dual overlap encoding. In Table 1 of [127], for ORFs > 200 bases in the *V. cholera* genome, the percentage with dual encoding is 6.57% (dual occurs when overlapping opposite read directions, with scores like 11000, 12000, and 21000). For Deinococcus radiodurans the amount of dual encoding is 69.4%.

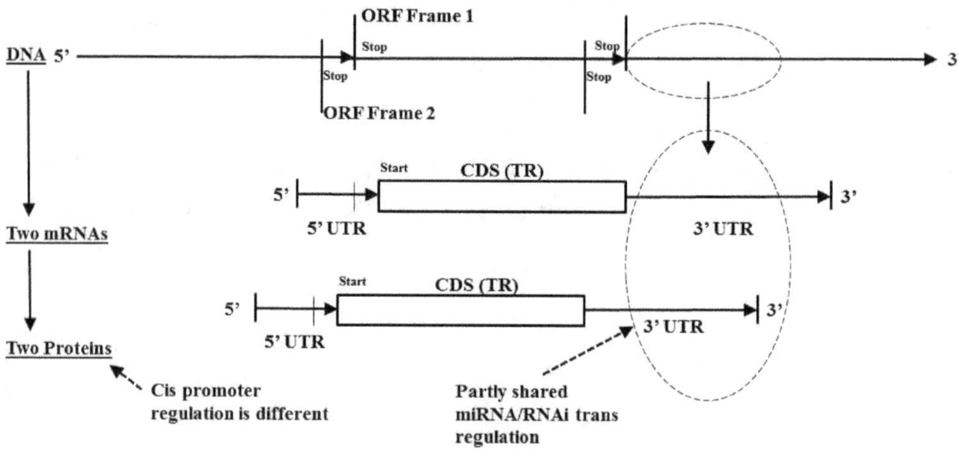

Figure B.2. mRNA Transcriptome Topology: Overlapping monicistronic

Once the prokaryotic transcription to mRNA is complete a selection for the RNA-based coding region is effectively done, and unique protein products are thereby directly indicated. There can be multiple protein products because prokaryotes can have polycistronic transcripts, whereby a single mRNA may have multiple, sequentially located (non-overlapping), regions that each encode their own protein product (a.k.a, operons), while eukaryotic transcript processing is more complex due to introns and alternative splicing. Once a eukaryote reaches the same stage of having a 'mature' mRNA, on the other hand, the resulting encoding is typically monocistronic (with cis-regulation governing only one encoding region). Simple eukaryotes, such as *C. elegans*, are known to have both types of encoding (monocistronic with introns and polycistronic without introns) in significant numbers.

The process of DNA→mRNA→Protein production is regulated at both transcription and translation polymerase stages (see Fig. B.3). *Cis* regulation dominates at the DNA→mRNA polymerase stage, and *trans*-regulation at the mRNA→Protein polypeptide production stage. In the case of the polycistronic encodings, there is one *cis*-regulatory region for multiple coding regions. The dominance of trans-regulatory mechanisms at the mRNA→protein stage is significant because all living processes, including viral processes, can be regulated at this stage, and many of the regulatory processes involve simple antisense nucleic acid molecular recognition, indicating a possible common and ancient (RNA World) biomolecular process. In eukaryotes the process of DNA→mRNA→Protein production is also regulated at the spliceosome level (for which a brief background is given in the next section). The main mechanism for *trans* regulation in eukaryotes is RNAi. The role of *trans*-regulation in prokaryotes involves a non-RNAi process *that employs no RdRp for siRNA amplification*, using a method evidently separately evolved: CRISPR/cas.

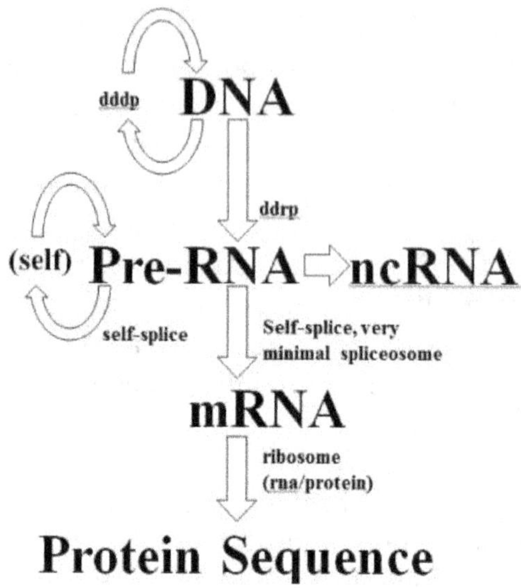

Figure B.3. Standard Central Dogma for Prokaryotic Cells. Minimal splicesome indicated, zero for many. Early cells eventually lose spliceosome and RdRp, but still have self-splicing introns to some extent.

B.2 Mobile Nucleic Information Structures

Viruses are a well-known exception to the central dogma and have already modified the central dogma when it comes to the role of reverse transcriptase (see Fig. B.4). Using reverse transcriptase, as the name suggests, it is possible to go back from RNA to DNA, altering the cell's genomic repository of information (in an inheritable way even in multicellular eukaryotes if that genomic DNA happens to be in a gamete cell). Viruses also have their own means for producing (or replicating) RNA information by way of the aforementioned RNA-dependent RNA polymerase (RdRp), while the production of pre-mRNA from DNA in the central dogma is by way of a DNA-dependent RNA polymerase (DdRp). The viral information processing, thus, includes everything in the central dogma up to the protein-production (cytosol) stage, as shown in Fig. B.4.

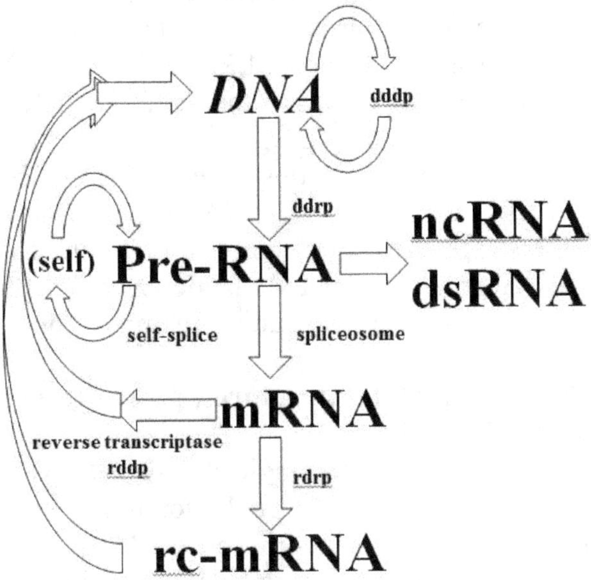

Figure B.4. Central Dogma for Viruses (have RdRp). Virus has minimal cytosol, no ribosome, no protein production, since this is taken from cell, as is much of the DNA processing. Virus retains RdRp and reverse transcriptase (RdDp, DdDp), RdRp, (so nucleic acid polymerases), and spliceosome.

The typical viral info-flow is RNA → Protein initially, then RNA→mRNA→Protein (capsid). The genomic diversity among viruses, however, appears to have explored the DNA/RNA encoding space in every way conceivable, and that's from the 5000 viral genomes used in an initial study, with hundreds of millions of viral genomes unknown. Genomes can be made from DNA, RNA, or both at different stages in the life cycle. The genomes topology can linear or circular, or segmented into linear and circular components. The DNA or RNA can be single-stranded (ss), double-stranded (ds), or partly both. The sense of translation can either be positive (direct encoding from base sequence, where positive sequence is what is recorded in the genomic databases, unless specified otherwise, by convention), or negative (direct encoding off of RNA-pairing product, e.g., the reverse strand read direction), or ambisense (some ssDNA and ssRNA viruses have some transcription off of the negative strand in their double-stranded replicative intermediates – the geminiviruses, for example).

The smallest virus has 2 proteins and is 2 kilobases in length (circoviridae). One of the largest virus has more than 1,000 proteins and is over 1.2 Mb in length (mimiviruses). All RNA viruses use their own RNA replicase enzyme to create copies of their genome. RNA viruses generally have smaller genomes than DNA viruses. This is consistent with the higher error rate when replicating based off of a RNA template (whether ssRNA or dsRNA), or even a ssDNA template. Stability of dsDNA genomes allows less error, with corresponding increase in genome size allowed. Some evolutionary adaptions were used by the RNA-based genomes to off-set this, including segmentation.

B.3 Cellular Nucleic Information Structures, prokaryotic

B.3.1 prokaryotic cellular genomes

Prokaryotic Genomes can be ssDNA (single stranded DNA) or dsDNA (double-stranded DNA), with linear or circular topology, and possibly segmented (segments referred to as chromosomes). Chromosomes usually only have one or two parts (with the second often only a plasmid). Genes are transcribed from the 5' to the 3' ends by the molecular machinery, where the region preceding the 5' end is the '*cis*' regulatory regions, and the region after the 3' end is the '*trans*' regulatory region. In prokaryotes one coding region maps to one protein product (see Fig. B.5).

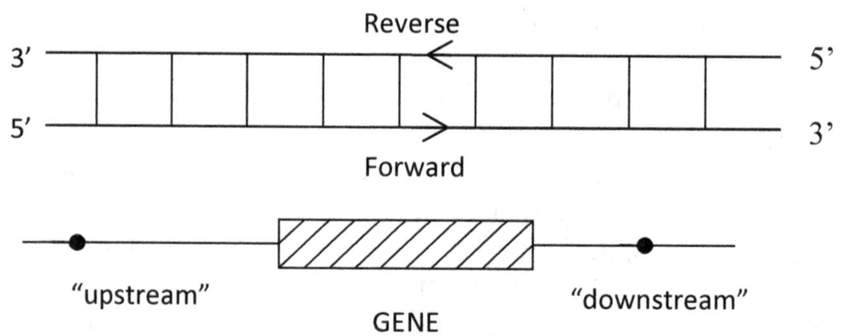

Fig. B.5. Prokaryotic Genome. Top is the Duplex DNA in a 'ladder' diagram, where the rungs of the ladder mark the hydrogen bonding between Watson Crick base-pairing between the two strands. **Bottom** is the standard one gene region one gene product structure (the initial hypothesis of gene encoding).

The typical prokaryotic coding sequence is shown in Fig. B.6. The typical gene constructs are: protein coding; RNA specifying; un-transcribed; and pseudo-genes. The core promoter zone is shown in Fig. B.6 as tatawaw (note: y= c or t; w = a or t; and r = a or g). The consensus Shine-Dalgarno sequence, aaggaa, is shown preceding the transcription start site (TSS), where mRNA transcription begins. Approximately 10 bases of the 5' untranslated region (5' UTR) separate the TSS from the start of coding, with possible presence of secondary (hairpin) structure that is inhibitory. The start of protein coding commences with the atg codon, and the end of protein coding is at the codon prior to the stop codon taa (the taa codes for stop, not for an AA). The region from codon to end of the mRNA transcript is referred to as the 3' UTR.

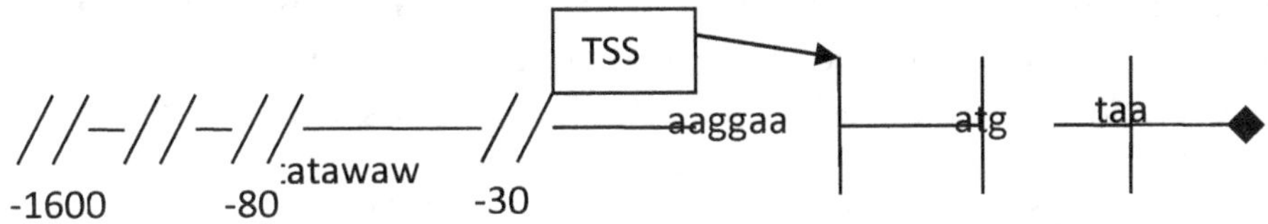

Fig. B.6. Prokaryotic start-of-coding region.

In prokaryotic genomes genes are often linked under common regulatory control in groups called operons: Common Start – TATA box – TSS – G1 start – G1 stop – G2 start – G2 stop – G3 start – G3 stop – Common Stop. Fig. B.7 shows such an operon, where individual genes can overlap in their encodings.

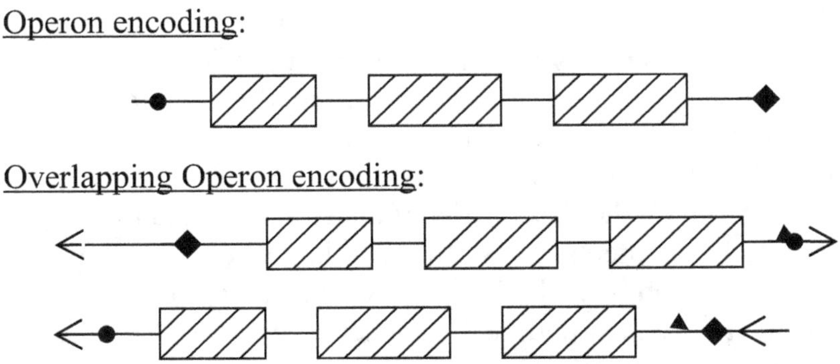

Fig. B.7. Prokaryotic gene structure.

B.3.2 Two types of membrane

Lipid/proto-metabolic World settles on two cellular membrane compartmentalizations, but with the same nucleic acid encoding scheme and primitive processing. How is this possible unless RNA World precedes the two cellular membrane compartmentalizations (e.g., the non-cellular phase of RNA World)?

B.4 Early transcriptomes and emergence of RNAi & CRISPRi

Although RNAi has been discovered only recently (ca. 1998 [464] and 2001 [465]), and is thought to have only become significant with the introduction of eukaryotes, some aspects of RNAi appear to be ancient and even fit within the RNA world paradigm. RNAi appears to be a universal process used by eukaryotes and their viruses, but not prokaryotes. A key component of the RNAi process is the use of a miRNA that is incorporated into a collection of argonaute proteins in what is known as the RISC complex [465]. The miRNA provides the nucleic acid sequence template that is used in regulating specific antisense-related mRNAs (see Fig. B.8). The miRNA's guide strand provides an antisense match to its mRNA target and plays a role in regulation or complete inhibition (if the target ssRNA is viral or a transposon). miRNAs may have had an early (RNA World) role as siRNAs given the discovery of a simple biogenesis pathway for miRNAs: 'mirtrons' are introns that once spliced anneal to themselves to form a miRNA. Mirtrons and mirtron-based RNA interference are also consistent with a "follow thc introns" analysis that suggests that introns might have been fundamental and ancient. The appearance of RNAi throughout the eukaryotes is an indication of its ancient eukaryotic origin.

RNAi is an RNA interference method that requires formation of a dsRNA intermediate. In the ssRNA gene-silencing experiments in 1998 [464] it was found that dsRNA was not as effective as ssRNA at silencing, and that the ssRNA silencing was "too good" in that a small amount of ssRNA would accomplish complete silencing, indicating an enzymatically catalyzed silencing process. The RNAi process was clarified in 2001 [465] with the description of Dicer and its role in preparing 22bp dsRNA segments for ssRNA template loading into the RNA-induced silencing complex (RISC). RNAi interferes with specificity by using a ssRNA strand that is antisense to the ssRNA target of interest (a mRNA or a retrotransposon RNA intermediate, for example). The current form of RNAi may not have been around with the early eukaryotes. Early eukaryotes may have had a proto-RNAi that was only mirtron based, and may not have had a significant RISC complex developed yet.

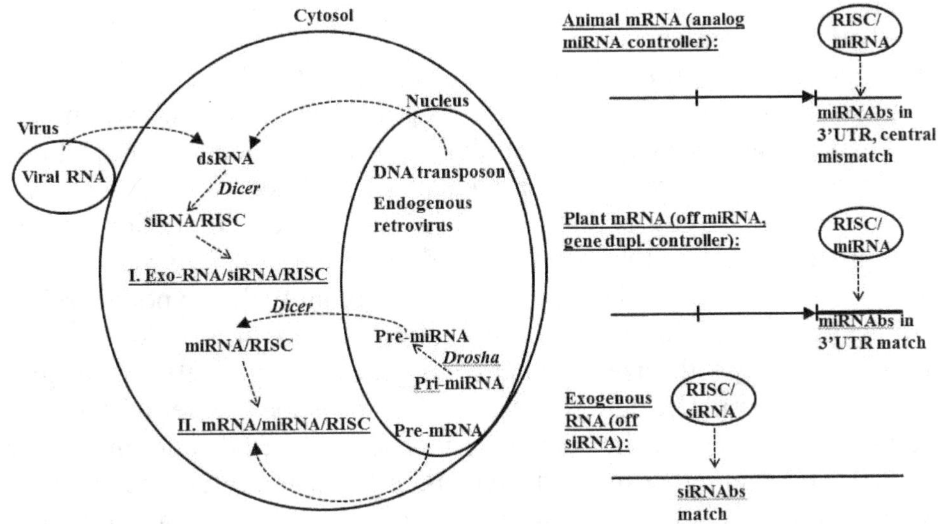

Figure. B.8. RNA interference (RNAi) defense and expression regulation in eukaryotes. Drosha clips pri-miRNA from mRNA form to the pre-miRNA form for export to Dicer. Dicer targets dsRNA intermediate (transposon , viral, and pre-miRNA) and creates miRNA or siRNA template and recruits RISC complex to hold ssRNA specific-binding template. When RISC complex with template binds RNA target have cleavage or binding according to specificity/stability of template match. A central miRNA template and miRNA binding site (miRNAbs) mismatch leads to gene downregulation, while a perfect match results in a gene or ncRNA knockout. Strong binding site match region is 7-8 bases in length. Strong match, length of match, vs mismatch…Epigenetic machinery of cell is partly regulated by a subgroup of miRNAs known as epi-miRNAs [466].

In the viral eukaryogenesis hypothesis, to be discussed in the next section, we see that an RNAi interference CRISPR/cas precursor is hypothesized to have existed in the archaeon/prokaryote-like cell (possibly with mitochondria) when it is invaded by a membrane-bound viral endosymbiont, where the membrane-bound invasion of the virus involves an RNA interference arms race (whereby the viruses membrane can protect and complete critical mRNA processing prior to release into the prokaryote-like host's cytosol). Thus, the virus endosymbiont may have resulted from a mutually beneficial symbiosis (mutualism) on the basis of RdRp usage and RNAi co-evolution, that eventually resulted in the wholesale adoption of the viral machinery and separate nucleus processing in the hypothesized viral eukaryogenesis.

B.5 Viral Eukaryogenesis

B.5.1 The Last Eukaryotic Common Ancestor (LECA)

The possible ancient viral origin of RNA polymerases is described in [467], where there is a hypothesized loss of RdRp in the ancient cell line that is regained from virus in later eukaryotic cell lines for cytosol RNAi support (via siRNA production). It's possible that the introduction of RdRp processes into both the cytosol and the nucleus of the proto-eukaryotic cell line was via the viral endosymbiont that is hypothesized to form the proto-eukaryotic nucleus, e.g., RdRp's adoption could be a remnant of the hypothesized viral eukaryogenesis event itself. The origin of many critical nucleic acid processing enzymes, the transfer of nucleic acid processing methods in particular, appears to have been dominated by viruses transferring their methods to cellular hosts and rarely the other way around [468]. Perhaps the spliceosome is viral in origin as well, or at least co-evolved with an ancient archaeon/prokaryote host. If the latter, however, we would expect to see spliceosomal activity in archaeons/prokaryotes, but there is no evidence of such. So simpler is to hypothesize that the list of hypothesized nucleic acid processing methods of viral origin, like RdRp, mRNA capping, and mRNA polyadenylation, also includes spliceosomal machinery. Details in support of this include *spliceosomal* introns (e.g., not the self-splicing Group I and II introns) are not seen in prokaryotes or archaeons, but are seen in viruses large enough to bother with the complexity [469], where the megaviridae class of viruses includes viral genomes larger than some cellular genomes [469-471], and in *Mollivirus* (infects *Acanthamoeba*), for example, has about 4% of its genes with *spliceosomal* introns.

Viral-based biomolecular machinery for passing virus molecules, including the viral genome, into a target nucleus is well-studied as it is a fundamental trait of all eukaryotic viruses [472-478] – the exception being those viruses, usually very small, that attack during mitosis when the nuclear membrane is temporarily disassembled (the parvo virus, for example, is only about 5,000 bases long) [472-478]. So, viral machinery for nuclear transport is well known and it could have been prevalent and co-opted in a mutualism context, where a host cell having multiple large viruses and multiple prokaryotic endosymbionts could have existed in a proto-eukaryote then as it is does in amoeba now. Viral control of access to the present-day eukaryotic nuclear envelope [472-478], or possibly the nuclear-like envelope of another virus in a large cellular host [239], and viral control of the spliceosome [479,480], and viral encoding of spliceosomal molecules [469,481] for optimized processing on viral encoded genes that require spliceosomal processing.[469,481], all lay the groundwork for a possible viral origin for the spliceosome in eukaryotic cells.

The standard hypothesis for the origin of the spliceosome in non-viral eukaryogenesis hypotheses is that it evolved from self-splicing (Group I and II) introns imported from the prokaryotic endosymbionts. And such a non-viral origin for the spliceosome has been suggested in the viral eukaryogenesis context as well [482]. The Results presented here appear to favor a *viral* origin for the spliceosome via the viral eukaryogenesis hypothesis. Results are shown that indicate an operational spliceosome in the proto-nuclear envelope as non-nuclear genomic information was being 'adopted', e.g., the spliceosome was under control of the proto-nucleus during the proto-eukaryote's uptake of the non-nuclear (archaeon and prokaryotic) genomic material. The spliceosome likely evolved from the Group I and II self-splicing introns, as suggested by others [482], but this spliceosome development may have occurred at a much earlier time than viral eukaryogenesis. This is hypothesized to be the case given the sizable percentage of the eukaryotic transcriptomes' (5% to 15% of mRNA *transcripts*) that have overlapped encodings that represent coding artifacts that are 'intron-less' in origin, e.g., such as would occur with an archaeon/prokaryote ancestor having dually encoded transcriptome material (imprinted from dually-encoded intron-less genome). Some of the overlap encoding artifacts even appear to be functional given their long ORF encodings (greater than 300 bases) and long 3'UTR regions (greater than 200 bases).

The possible viral origin of the spliceosome, which is beginning to be indicated by results like those in [469], and described here, is relevant to two hypotheses: "Virus World" and "Introns Earliest". The Introns Earliest Hypothesis was mentioned in Ch. 5, so will only be mentioned briefly in what follows. The Virus World Hypothesis suggests that the role of viruses could be significantly underestimated by the standard descriptions. In [483] it is even suggested that DNA and DNA replication may have first appeared in Virus World, from which three RNA cell lineages, each with its own symbiotic DNA virus lineage, could have then led to the cellular domains seen today [484].

Pre-eukaryotes may have co-evolved an early form of RNAi (typical archaeons/prokaryotes have an unrelated form of RNA interference via

CRISPR as described in Sec. 8.4) with a mega Virus, where the battle for RNAi control of host, eventually gave rise to the eukaryotic cell via modulation of RNAi. If nothing else, RdRp's have a pervasive role in eukaryotes in defense against foreign nucleic acids, in developmental regulation, in genome maintenance, and in transcriptome modulation [485]. The RNAi transcriptional and post-transcriptional silencing in *Schizosaccharomyces pombe*, for example, is accomplished with a single version of Dicer, RdRp, and Argonaute protein (Ago1) [486]. Some theories of viral eukaryogenesis indicate that the development of RNAi was a critical escalation in the battle for cellular control that led to the viral-nucleus dominated viral eukaryogenesis pathway [467]. Viral eukaryogenesis has been proposed in a variety of forms [467,487-489, 505], all consistent with the lack of RdRp in prokaryotes/archaea (as with RNAi), while RdRp is found in eukaryotes and viruses (as is RNAi). Viral eukaryogenesis is also indicated by a number of other recently discovered structural properties, including the ease of membrane vesicle bound transfer of genomic information [490], and an emerging picture of virus complexity and their co-evolving relation with cells [491-495]. So the important role of RdRp for RNAi purposes clearly explains why RdRp would have singular importance to eukaryotes, but is there a non-RNAi role for RdRp as well? In some RdRp RNA productions lengthy anti-sense strands are produced, and this suggests a possible non-RNAi role, as will be discussed later.

As briefly discussed in Sec. 5.5.3, the viral endosymbiosis could have been a very slow transitional process given the likely mutualism 'melting pot' of endosymbionts, including both proteobacteria (eventually leading to mitochondria, etc.) and viruses (including mega viruses) that could have existed at the LECA. This could have existed then since it still exists now, in present day *Acanthamoeba*, which is the host to the megaviruses Mimivirus, Marseillesvirus, and other nucleocytoplasmic large DNA viruses (the NCLDVs), as well as a variety of bacteria [496], often hosting multiple virus and bacterial visitors at the same time. Furthermore, phylogenetic analysis shows that the NCLDV viruses have an ancient relation to eukaryotes, thus the LECA [497-500], placing them in the mutualism context at the time of the hypothesized viral eukaryogenesis. The NCLDV viruses possess their own mRNA capping protein (eIF4E), mRNA capping enzyme, and DdRp, to ensure preferential production of their mRNAs by the host, and eIF4E, at least, is thought to be unique to the virus (not obtained from the host *Acanthamoeba*) [501-503]. The mutualism 'melting pot' environment of the early eukaryote formation is further supported by viral mutualism in the form of favorable RdRp exchange from T7 to the mitochondrial host genome, which is thought to have occurred at the time of the LECA (or earlier) since it is present in all known eukaryotic mitochondria (with only one known exception [496]).

The viral eukaryogenesis theory leaves little opportunity for testing since it relates to a hypothesized ancient historical event, and for this reason testing the viral eukaryogenesis theory has been grouped with other hypothesized endosymbiosis events, such as those endosymbiosis events leading to the mitochondria and chloroplasts [504]. The viral eukaryogenesis hypothesis, however, involves a shift of non-eukaryotic cytosol based genomic data into the newly formed nucleus of the organism, so may leave more evolutionary artifacts in the genomic sequence information than could be obtained in the case of the organelles (see Sec. 8.6 & 8.7).

Antisense encoded mRNA information is known to exist. In Sec. 8.7, however, we see anomalous amounts of *overlapping* antisense encoding, with overlapped transcriptome encoding at percentages seen in some prokaryotes and archaea. Even if the antisense encoding aren't functional mRNAs in their own right, they are artifacts of such, and in sufficient numbers to indicate the non-eukaryotic (non-spliceosomal) genomic information was imported into the proto-eukaryotic nucleus by way of transcriptome adoption, apparently favoring uptake of a dominant transcript. If any of the antisense encodings are functional this would then require use of an RdRp. Such use of RdRp is already well known to exist, however, for shorter, partial transcripts, where it is used to get siRNA for RNAi regulation and control. Thus a role for RdRp that is not RNAi related is suggested, in order to access "ghost" antisense transcripts from the ancient prokaryotic/archaeal ancestor, and to allow for the overlap encoding in general.

To understand the nature of the imprint artifact from the prokaryotic/archaeal genome/transcriptome to the eukaryotic transcriptomes, it helps to first review some background on genome and transcriptome coding topology), and overall classic central dogma for archaeons/prokaryotes. Since archaeons/prokaryotes very rarely exhibit introns (some do have self-splicing introns, however, and also for some viruses too [505]) their transcriptome structure almost directly maps to their genomic structure.

B.5.2 Viral Eukaryogenesis Hypothesis

Given the possibly widespread and thus ancient use of RdRp in eukaryotes, this would indicate that the inception of the proto-eukaryotic cell line would have been marked by the adoption of a native RdRp capability. Consider also that at the inception of the eukaryotic cell line there is the adoption of a nucleus (by definition) with the possible wholesale adoption of a spliceosomal machine protected by the nuclear membrane, as well as any other nucleic acid processing that might be protected by that membrane boundary. For the proto-eukaryote there is also thought to be a more energy-rich metabolic suite of capabilities than is typical for prokaryotes. For this reason the archaeon/prokaryotic-like organism involved in the viral eukaryogenesis may have already adopted a proto-mitochondrial endosymbiont, allowing it to perform significantly more ATP production at the mitochondrial membrane, not the cellular membrane, freeing the cellular membrane to become simpler (e.g., no wall, just a single membrane) allowing greater cellular communication (and greater susceptibility to membrane-bound virus invasion). Now consider the archaeon/prokaryotic-like cell scenario in the co-evolutionary context of an ancient evolutionary battle between virus and cell. As the archaeon/prokaryotic-like cell develops more and more refined RNA interference defenses it forces the co-evolving virus to compensate, by learning how to co-opt those defenses for its own use, and by adoption of a membrane boundary for invasion by internalization and operation as an internal (nuclear) boundary (in some cases, such as for poxviruses [487]) to shield its RNA processing from interference. All that's needed for the host virus system to move towards mutualism is a trade-off, such as the host getting the use of RdRp, where the virus gets the usual access to the host cellular machinery. Viral eukaryogenesis is, thus, hypothesized in this picture as the simplest description given the organisms and their interplay at the time.

A remarkable 'fingerprint' will be shown from the prokaryotic-like precursor's genome/transcriptome in the eukaryotic transcriptomes we see today. This is because the retention of the archaeon/prokaryotic-like genomic information seems to be by way of adoption of the archaeon/prokaryotic-like organism's transcripts, where the transcript that carried a particular gene with an overlap encoding is adopted, with overlap information intact, into the proto-eukaryotic transcriptome processing, and eventually (possibly via reverse transcriptase) gets written into the viral/nuclear dsDNA genome (assuming a dsDNA virus).

Thus, the highly compact overlap, operonic, and *dual* encoding found for prokaryotic genes is found imprinted not via genome-genome transfer, but via transcriptome-genome transfer. Given the splicing to arrive at transcript, however, this results in the transcript only producing genomic information from one strand, with the other (dual) information necessarily accessed via RdRp. With the viral mutualism hypothesized already in-place, however, this accessibility of RdRp would have not been a problem.

There are a number of assumptions leading up the specific forms of viral eukaryogenesis hypothesis outlined above. Part of the 'stage that is set' involves the proto-archaeon/prokaryotic cell and ancient virus co-evolution in place at the time. Whether viruses had spliceosome processing already (and it was ancient) or whether nucleic acid splicing was a remarkably late invention, this is another assumption that impacts the viral eukaryogeneis model. In [505] and here, the Intron Earliest hypothesis is posed, such that the viruses arrived at the viral eukaryogenesis with splicesome already present, this being one of the selection pressures for them to protect against RNA interference by adoption of membrane enclosure in their 'trojan horse' endocytosis attack (that eventually leads to endosymbiosis when entering a mutualism relation). In the Intron Earliest hypothesis, the spliceosome processing is itself ancient and carried in the viral line since selection pressure on proto-prokaryotes led to loss of introns (and loss of the need for the spliceosome). RdRp is similarly thought to be ancient and carried in viral line and similarly lost in proto-prokaryotic line as shift to DNA and larger genomes (and host complexity) led to shift to DdRp in cells.

The viral eukaryogenesis theory also suggests a possible origin for meiosis and sex [504], where it is proposed that the mitotic cycle evolved from virus established with a permanent lysogenic presence, and the meiotic cycle (and sex) evolved from the process whereby the virus transferred to new hosts. Also in [504] is discussed the process whereby the viral-nucleus dominated cellular control led to a reorganization of the prokaryotic transcription/translation regime into the typical eukaryotic process whereby mRNA is capped prior to extrusion into the cytoplasm, and where the cap binding protein directs translation of the capped mRNA. Reorganization covers an evolutionary refinement process in the proto-eukaryote where non-nuclear genomic information gets 'lifted' into the nucleus. Perhaps the same proteins that participate in the spliceosomal activity in the nucleus, bind nucleic acid in cytosol on their way from their cytosol production and assembly into the nucleus [479,480], allowing reverse transport, with reverse splicing of introns already occasionally occurring inside the nucleus, reverse transcriptase (RT) would then be all that's needed to pull the 'intronified' transcript information permanently into the viral/nuclear genome. This could easily occur since RTs are a common component of viruses, even the simplest viruses. In the evolutionary lift procedure outlined, an archaeon/prokaryotic host transcript would be mapped to an 'intronified' viral-nucleus genome sequence. This is hypothesized to have occurred given the distinctive statistical artifact that is seen in Sec. 8.5 – e.g., eukaryotic mRNA transcripts are seen with overlap encoding information with respect to their reverse complement reads (tracing back to non-eukaryotic mRNA transcripts with such overlap encoding, which then relates directly to the same overlap encoding at the non-eukaryotic *intron-less* dsDNA genomic level). The results, thus, support a form of the viral eukaryogenesis hypothesis with a genomic uptake via 'intronification', further suggesting that the viral nucleus already had control of the spliceosomal machinery. This also suggests a possibly larger role for RdRp in eukaryotes than purely RNAi-related.

B.6 Cellular Nucleic Information Structures, eukaryotic

There are six chromosomes for Worm (*C. elegans*). As mentioned in Ch.5, for Eukaryotes the Gene → Protein translation is not collinear due to the existence of "introns," as is shown in Fig. B.9.

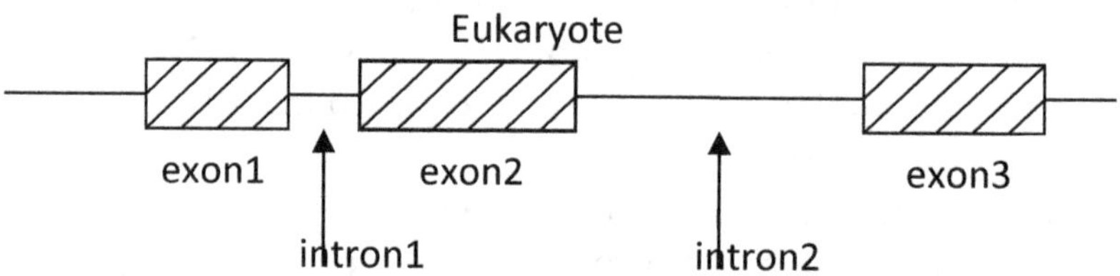

Fig. B.9. The Exon/Intron structure of a 3-exon eukaryotic gene.

The eukaryotic cell has two separated sets of machinery that work on the nucleic acid information, one to transcribe DNA to mRNA, and one to translate mRNA to protein, the overlapping imprint of structures (motifs) is more complex for this reason. The transcription factor (TF) binding sites (TFBSs) are usually located upstream of the gene (*cis*), and help to modulate the transcription of the downstream gene product. The cat box, tata box, and inr box, are shown in Fig. B.10, are motifs associated with the interaction with RNA polymerase (promoter signals). Also upstream are CpG methylation sites, where areas rich in GC are modifiable to "turn off" transcription. When the mRNA is transcribed it is given a 5'CAP at position -1. In the 5' UTR, there is typically a motif, the Kozac sequence (GCC G/A CC | ATG G), that is a ribosome binding site.

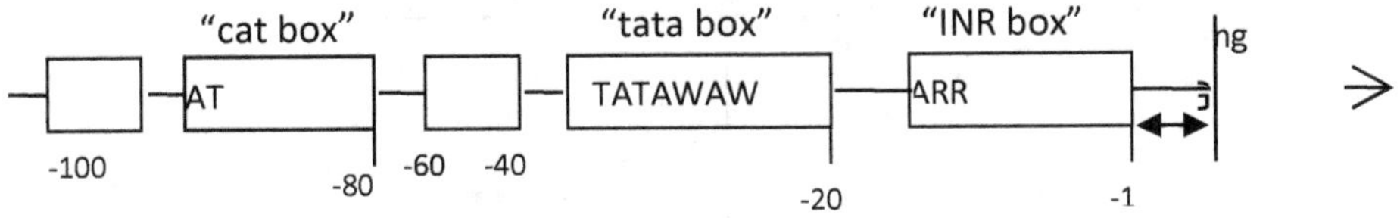

Fig. B.10. Eukaryotic coding region: the start of coding.

After the start of coding, eukaryotic genes have introns interspersed with incomplete coding elements (exons). There are three types of exons: first, internal, and last (see Fig. B.11).

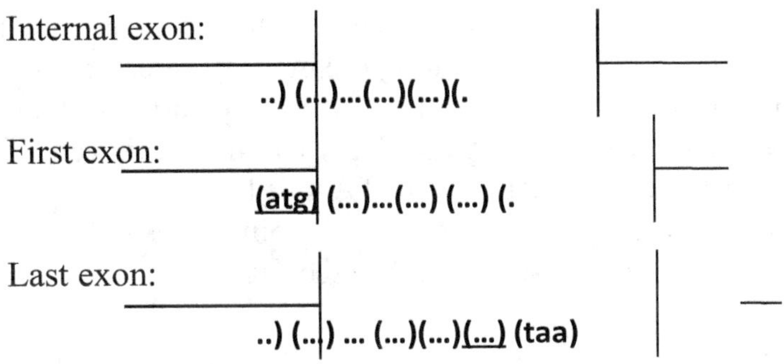

Fig. B.11. Exons: first, internal, and last.

The ends of introns are highly conserved and some are self-splicing (see Fig. B.12). Single Nucleotide Polymorphisms (SNPs) usually have only one variant → thus binary (on/off). If in promoter or protein coding, the SNP is very important, if in the intron region, not generally as important (unless at a splice site). The internal schematic for a generic intron is shown in Fig. B.12 (Y=C or T; N=anything).

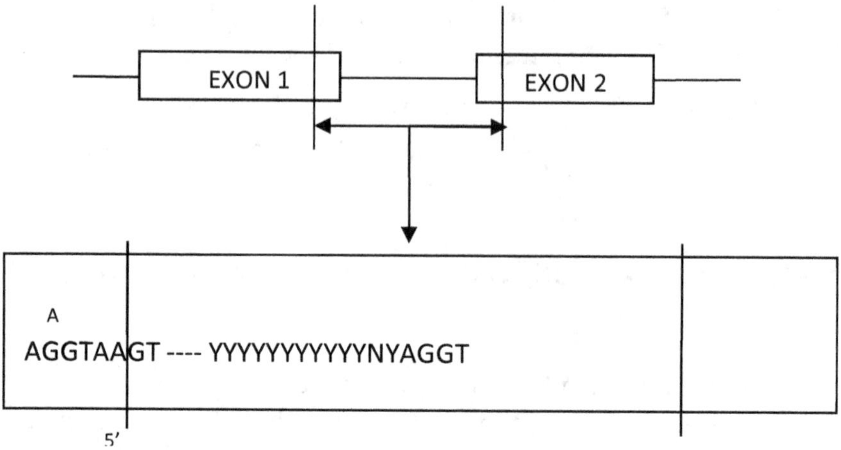

Fig. B.12. Intron Splice Signals

The non-self-splicing variety of introns have further artifacts indicative of their co-evolution with an enzyme. Here, the motif of interest is known as the branch point, or the adenosine moiety, which typically has length 19-37 nt, with consensus sequence NYTRAY, where the 'A' is very highly preserved. In yeast the consensus is TACTAAC. The branch point is part of the lariat loop (end stuck to itself at branch point) transient complexation/splicing by the spliceosome (see Fig. B.13).

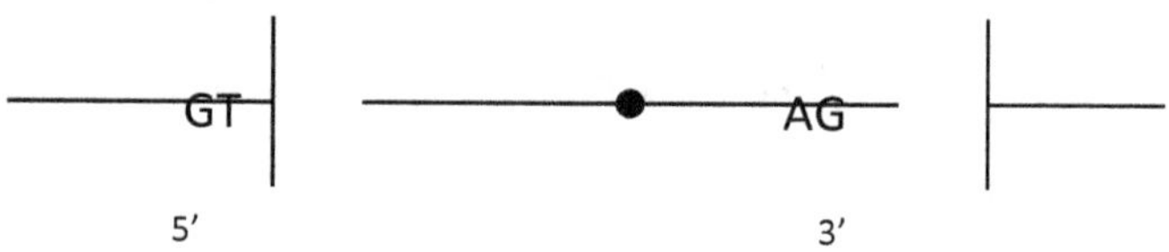

Fig. B.13. Intron with branch point marked.

To understand the end-of-gene encoding we must backtrack to the transcription stage, where the pre-mRNA → mRNA is complete with cleavage at the polyadenylation site is followed with attachment of the poly-A tail. The poly-A cleavage site typically has a motif AATAAA that interacts with the cleavage polyadenylation specifying factor (CPSF) followed by a CPSF stimulation factor (CstF) consisting of 15-25 nucleotides of GT rich region trans to the CPSF motif. The distance from the end of coding to the CPSF can be quite large, on the order of 100 nucleotides.

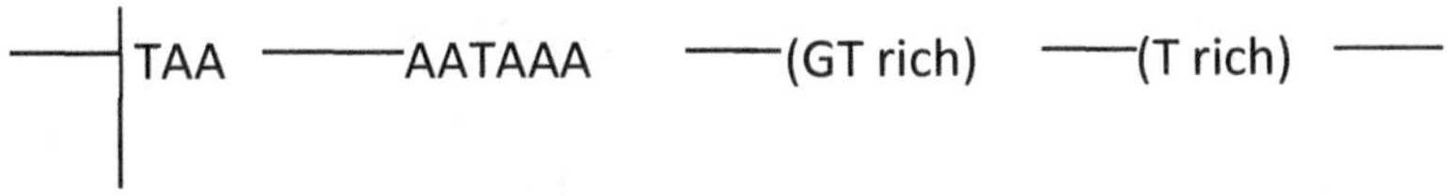

Fig. B.14. Eukaryotic end-of-coding structure.

It is found that there is a preserved order of exons (synteny) between closely related species. This permits comparative genomics at the mammalian level (see Fig. B.15), with most differences in the length of the introns.

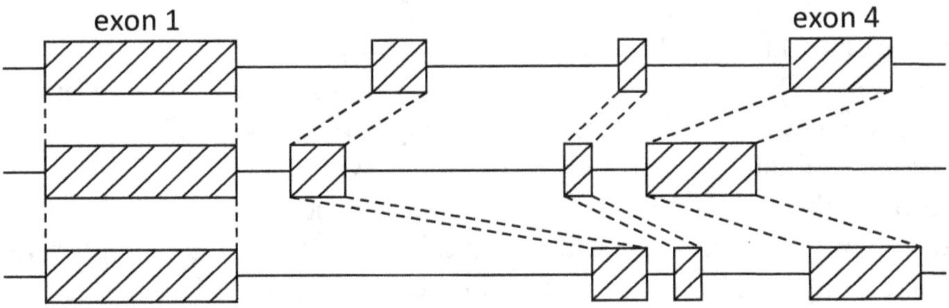

Fig. B.15. Synteny offers the ability to do extensive comparative genomics at mammalian level. Introns (and their lengths) can be variable while exons are more conserved. Coding exons are ≈ 150 bases, while introns are ≈ 500 bases (with some as large as 100,000 bases or more).

The eukaryote *C. elegans* (*Caenorhabditis elegans*) has a genome consisting of ≈ 97 million base-pairs of DNA (≈ $^1/_{30}$ human, ≈ 8x yeast). C. elegans has 6 chromosomes, denoted: I, II, III, IV, V, X (roman numerals), and its chromosomes have cis- and trans-splicing as well as operon structure (it still has prokaryotic genomic remnants…). An individual gene construct for *C. elegans* is shown in Fig. B.16. Note the existence of the 'outron' construct for *C. elegans*, essentially an intron on the outside, in front, with a 3' consensus at the start of coding end, but no 5' consensus prior to the INR (in the resulting 5' UTR). Non-gene repeats are thought to aid in signaling at splicing, at borders between AT and GC rich regions.

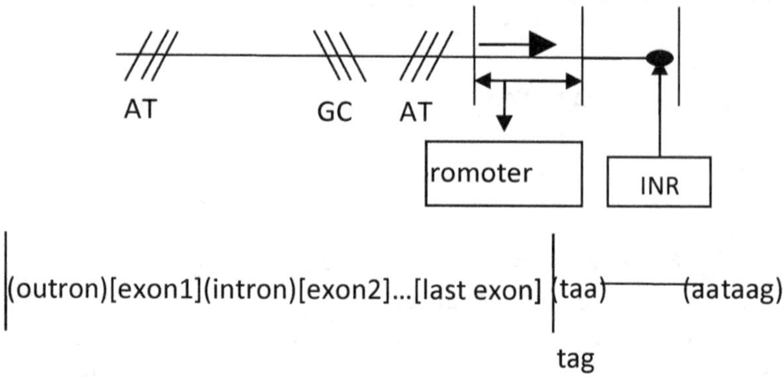

Fig. B.16. ***C.elegans*** **gene/outron structure.**

B.7 Transcriptomes and their Co-regulation [127]

The process of DNA→mRNA→Protein production is regulated at both transcription and translation polymerase stages (see Fig. B.17). *Cis* regulation dominates at the DNA→mRNA polymerase stage, and *trans*-regulation at the mRNA→Protein polypeptide production stage. In the case of the polycistronic encodings, there is one *cis*-regulatory region for multiple coding regions (as indicated in the first section). The dominance of *trans*-regulatory mechanisms at the mRNA→protein stage is significant because all living processes, including viral processes, can be regulated at this stage, and many of the regulatory processes involve simple antisense nucleic acid molecular recognition, indicating a possible common and ancient (RNA World) biomolecular process. In eukaryotes the process of DNA→mRNA→Protein production is also regulated at the spliceosome level (for which a brief background is given in the next section). The main mechanism for *trans* regulation in eukaryotes is RNAi. The role of *trans*-regulation in prokaryotes involves a non-RNAi process *that employs no RdRp for siRNA amplification*, using a method evidently separately evolved: CRISPR/cas. Fig. B.18 shows a modified Central Dogma with reverse transcriptase and RdRp accessible to the system.

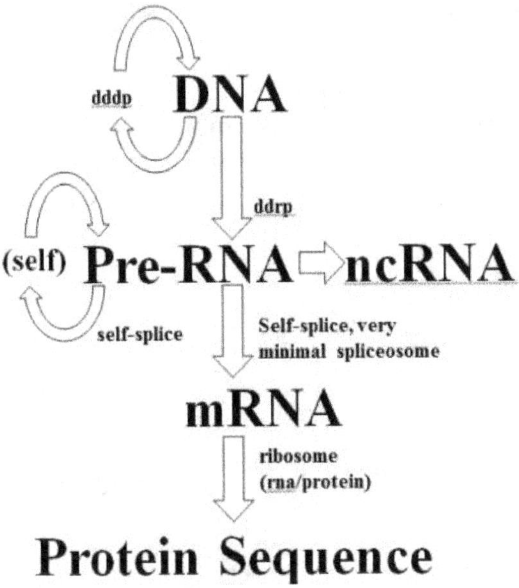

Figure B.17. Standard Central Dogma for Prokaryotic Cells. Minimal splicesome indicated, zero for many. Early cells eventually lose spliceosome and RdRp, but still have self-splicing introns to some extent.

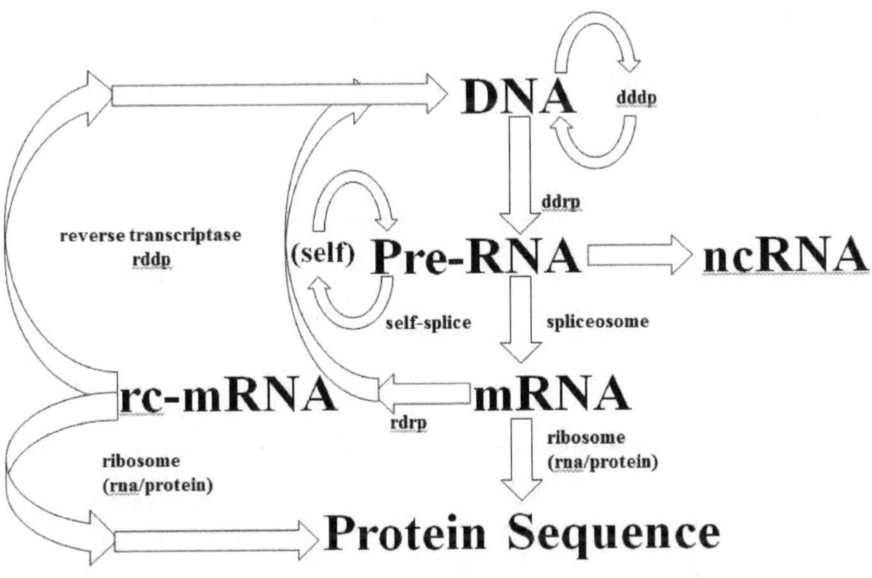

Figure B.18. Modified Central Dogma (with reverse transcriptase and RdRp). Shows the modified central dogma with both reverse transcriptase and RdRp present in the evolutionary process, early cells are hypothesized to have lost their spliceosome and RdRp processes to arrive at the familiar prokaryotic cell, as shown in Right.

B.7.1 From genome to transcriptome

The classic central dogma for the simplest, monicistronic, encoding was presented previously (and Fig. B.17). Here we have the modified central dogma (above Fig. B.18) and a diagram for the polycistronic case (Fig. B.19 & B.20 below):

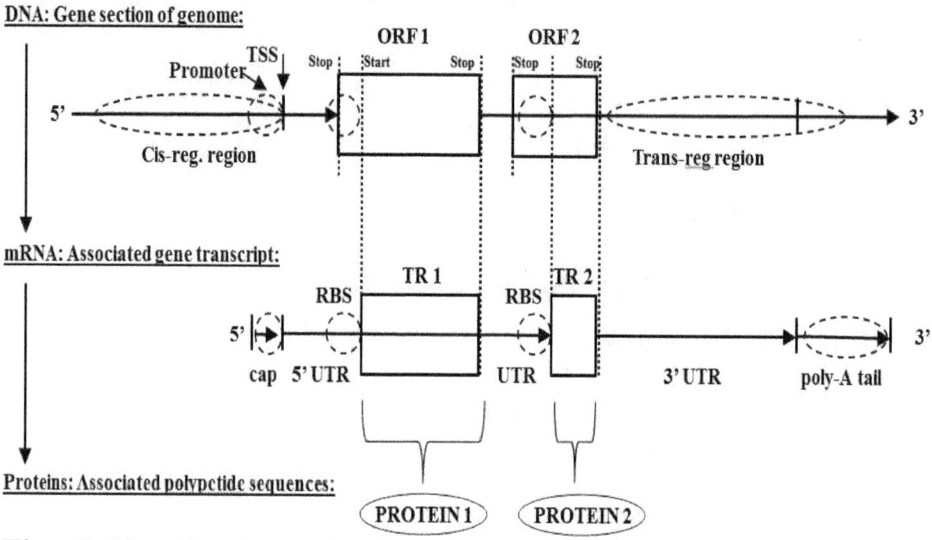

Fig. B.19. Classic Central Dogma, operon (polycistronic): DNA→mRNA→Protein Set.

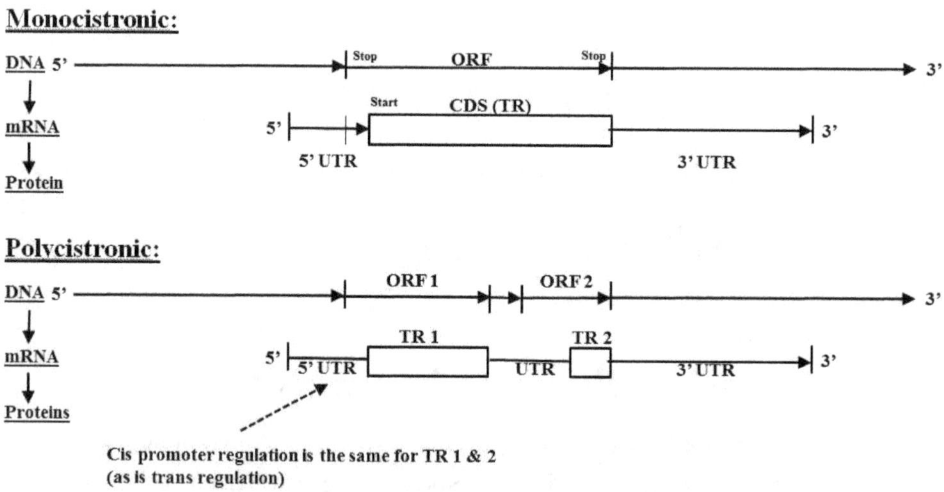

Fig. B.20. mRNA Transcriptome Topology: monocistronic and polycistronic

A long ORF can overlap with another ORF that has a different framing (since the codon has length three bases, there are three possible framings). When working with dsDNA genomes both strands encode, so with reference to one strand listed in genbank, say, one needs to perform three frame passes in the normal 'forward' direction, then three frame passes on the reverse complement of the reference strand, in order to identify all ORFs >= 300 bases. Once this is done it is found that many of the ORFs overlap. An accounting of this overlap 'topology' is indexed and plotted as shown in [127].

B.7.2 Transcriptome Analysis

Once the anomalous ORF structure is identified, nearby associated encoding anomalies are discovered (which in turn serve as validators), such as transcription start site recognition, in case of genomic sequence, or start/end of coding region recognition, in case of genomic or transcriptomic sequence information. The *cis*- and *trans*-regulatory regions are shown in [127], with *cis*-regulation via protein transcription factors dominating for DNA→mRNA regulation, and miRNA template strand recognition (via RNAi) regulation dominating mRNA→ protein processing.

B.8 Proteomics, Nanomachines, and Gene Circuits

B.8.1 Protein Nanomachines

The study of protein nanomachines as a new field of physics was suggested by Nobel Prize winner R.P. Feynman in lectures presented in 1959, and published shortly thereafter [506]. The main idea was that *individual* molecules might eventually be identified, observed, modulated, and utilized in a variety of ways. One of the most notable ways this has come to fruition is in the Nanopore detector described [505], and in the Nanoscope described in [507]. Also anticipated in Feynman's 1959 lectures was a new physics of bionanotechnology, starting from molecular 'nanomachines', but not made 'from scratch', rather from re-use of existing, biologically selected/optimized, protein nanomachines (as is accepted practice today).

Virtually all protein function involves specific binding to other molecules. The molecule bound by the protein is known as a 'ligand'. Anything can be the binding target or ligand, including small molecules, lipids, sugars, DNA, peptides, and other proteins. If the protein is an enzyme then the substrate is the ligand, for example.

Specific binding is possible due to the specific complementary pocket-like shape and interaction of the protein's 'active site' to its ligand target. This specific binding structure is critical to the functional role of the protein. Protein polymer is connected via peptide bonds. The peptide bond imparts critical rigidity and leads to the alpha helix and other stable conformational structures. A critical bond hybridization leads to planar peptide bond, leaving the polymer's conformational degrees of freedom largely simplified to the two orientation angles along the backbone at each C-alpha.

As mentioned previously, DNA encodes for the 20 'standard' amino acids, and a 21st (selenocysteine) and 22nd (pyrrolysine) are known to be in use as well (in simple, archaic, organisms). A protein is a polypeptide biomolecule in stable 3-D conformation. (Folding typically produces a globular shape, with a hydrophobic core and a surface that is hydrophilic dominated, sometime according to specific waters of hydration.) A polypeptide is any length, but implies lacking in defined 3-D conformation. A peptide is a short amino acid oligomer often lacking defined 3-D conformation. After translation, the residues of a protein are often modified by posttranslational modifications (glycations, etc.). After release of synthesized polypeptide, the molecule is selected for its ability to form a 3-D conformation with sufficient rigidity and surface attributes to have function.

It is interesting to note that a 22 amino acid alphabet is in use in encoding protein information, similar to a number of alphabets with 22 letters, where words based off of an alphabet with 22 letters are thought to be theoretically optimal for maximal information transmission in certain circumstances [1], so this is probably not a coincidence.

Although the emphasis on the protein structure-function relation is well warranted, it has also been over-used. In addition to molecular function, much of a protein's role is governed by location (cellular compartment) and timing (e.g., role in biological process). A protein of interest can be targeted with a chimeric molecule that includes (green) fluorescent protein, allowing the location of protein to be 'lit up' under fluorescent illumination. Protein that resides in the membrane will be clearly shown at the cell membranes, thereby outlining the boundaries of the cellular network like a Voronoi tessellation. Protein that is more distributed in the cytosol, or concentrated in the nucleus (if there is one) are other obvious compartmentalizations that could be revealed upon immunofluorescent protein tagging.

Ascertaining the function of protein can often best be done by 'knocking out' the protein and seeing what function is lost (the classic loss of function, with regain of function, family of experiments). Although highly informative, the individual nature of the knock-out analysis, as with the individual nature of the immunofluorescent tagging, leads to a process that won't scale well as the number of proteins of interest grows into the hundreds of thousands and beyond. In well-studied single-cellular yeast cells, however, there is an example of a partly-scalable knock-out analysis by use of transposons with a β-galactosidase payload. If the transposon successfully inserts at the start of a gene, it does so by knocking out the gene with an active payload. In this instance, the β-galactosidase, which if produced, turns a substrate blue, thereby indicating a mutant cell (with a knock-out) is present. The mutants can be isolated and grown (in colonies) such that genome sequence can be obtained, to reveal which gene, thus protein, has been knocked-out. Studies of the knockout strains are then done. This process is still immensely labor intensive, and requires some repetition, so is only partly scalable.

If we turn to the structure-function paradigm to help us with understanding protein function, we are again faced with an established approach, crystallography-based structure identification, that is non-scalable for large numbers of proteins. Worse than that, the approach isn't even scalable to proteins that are 'large' (greater than 200 amino acids in length). As the name suggests, the method requires the object of study to be convertible to a crystalline form. Therein lies the rub, since to achieve crystalline form requires a dominant conformer structure (the repeating configuration sub-unit of the crystal) that is solution soluble. This poses significant complication for any membrane-bound protein due to their hydrophobic components not being solution soluble, requiring detergents to be introduced to bind the hydrophobic parts allowing the molecule to be lifted into solution, preferably with a single conformer. Often molecules need to be broken and crystallized in parts, and this is even more typical (whether soluble protein or not) for proteins greater in size than 200 amino acids. The production of the crystals themselves can often take significant time (up to months, while for photorefractive crystals used in optics it's up to two years). For these reason, although the gold standard for their remarkable precision, the crystal-based diffraction methods do not scale for the genomic era.

Often a protein can have more than one structure, in which case the crystal analysis would need to identify situations where each structure dominated, to get a monoconformer collection for each structure, and thus get two crystal types for the diffraction analysis.

To get an idea how important a protein's choice of configuration is, with just two possibilities, say, consider the famous prions. What would happen if a protein has a conformation that induces other proteins of that type to take on that same conformation? – an infectious conformation as it were. If the protein shape thereby catalyzed also allows crystalline structures to form in vivo, then a disease condition results (Jakob-Creutzfeld in humans, bovine spongiform encephaly in cows, aka Mad cow disease). The discovery of prions as an infectious agent was quite a novelty, and the first discovery of an infectious agent that did not involve nucleic acid (Stanley Prusiner got the Nobel Prize for this). Prions as an infectious agent are fundamentally different for this reason (they can't be destroyed by heat, e.g. cooking, as in nucleic acid based infectious agent). The thought that a single prion molecule exposure could initiate a lethal crystallization (in brain tissues years later) is so fantastical as to be reminiscent of Kurt Vennegut's "Ice 9" (a humorous satire where water crystallization above freezing was sought, the Ice 9 conformer, to allow tanks to drive over mud frozen to hardness – in the end the Ice 9 gets out and freezes the whole world and everyone dies).

This is not to say the prions occurring in nature are entirely associated with disease conditions. There is a prion-like protein in yeast (Sup35p) that binds ribosome to induce loss of fidelity on recognizing stop codons. This results in a flurry of protein production that is non-standard, and in some cases appears to offer antibiotic resistance. The infectious protein is inheritable from mother cell to daughter cell, thus selection pressure reveals the benefits of Sup35p in such situations.

Enzymes
Cellular life requires a huge number of chemical reactions to occur, in a non-interfering manner, on timescales ranging from milliseconds to nanoseconds, where those chemical interactions would normally occur on the timescale of up to hundreds of years. This is accomplished by having virtually all biological reactions catalyzed. It has long been noted how

simple metals can aid in catalysis, but this occurs non-specifically (thus interfering), so what is required is an enormous collection of special catalysts for basically every biological reaction, where their catalytic activity is highly specific (thus non-interfering). This collection of catalysts is almost entirely protein-based (e.g., enzymes), but vestiges of nucleic acid based catalysts, indicative of an RNA World precursor, still remain (the ribosome being the most notable example).

The immense variety and complexity of modern protein-based catalysis is unlikely to have ever been matched by RNA World constructs. In fact, RNA World may have not allowed membrane channels to develop, significantly, due to their lack of hydrophobic domains (further details in the next sub-section). For this reason, the primary cellular expansion in the biosphere may not have occurred until in-step with the protein catalyst-takeover revolution, such that familiar basic cellular constructs, like channels, could be present.

The basic idea of a catalyst is that it lowers the potential barrier for a biological reaction to proceed (in a timely manner). It also needs to do this in manner that is non-destructive to the catalyst as well, such that a single catalyst can catalyze hundreds of interactions in the fraction of a second. But there's more to it than that. The rate of catalysis for an individual catalytic molecule is not found to be the same even in identical buffer, for which two main causes are attributed: (1) the exact same molecule can have slightly different conformational state, each with different catalytic rates; and (2) the same protein sequence molecule can have different post–translational modification, such as with glycosylation. In the case of (1), the catalyst can be modulated as to its catalytic strength according to secondary molecules (also protein or peptide) that serve to lock the catalyst into one conformation or another (with associated modulation of catalytic activity).

Proteins, such as enzymes, can have a high degree of variability. It has been demonstrated that enzyme turnover rate, for example, can differ at the single molecule level [508-510], with a single enzyme observed with one constant turnover rate, while another enzyme, differing only in conformation, or possibly by a difference in glycation, has a different, but still constant, substrate processing rate. And this is a simple example where there is only one interaction region and it is (mostly) unchanging in its conformation for the individual protein examined. Some allosteric proteins, on the other hand, with multiple binding sites for a particular target, change their binding affinity according to how many ligands they have bound. Antibodies are known to change conformation during binding to one (or two) antigens in such a significant manner that this is the basis for activation of the complement cascade of the adaptive immune response.

Channels and Transport Proteins
As mentioned in the prior section, it's hard to imagine cellular life really taking off without a rich assortment of membrane channels and transporters with which to regulate the chemical composition of the cell. In practice, proteins embedded in the cells plasma membrane are found to operate in a very highly regulated manner (where most variations are associated with a disease condition). Some proteins provide very narrow channels, allowing highly selective transport even on small ions. Channels even exist just for transport of water molecules (aquaporins) in a highly controlled 'ratchet' process that prevents a capillary entrainment catastrophe on water-transport. Channels exist for transport of specific larger molecules also, such as maltoporin. If transport can occur down an electro-chemical gradient, e.g. unaided, then it's referred to as passive transport via the indicated channel or transport mechanism. If energy is required to transport the molecule across the cell membrane, it's referred to as active transport. The standard supply of energy via ATP is used form many active transport situations. Sometimes, however, the transporter molecule, when bound to transport target, provides an autocatalytic transport activity.

The typical membrane channel formation is an ATP driven process. In the case of cellular pathogens, however, their pore-forming toxins would not be expected to knock on the cells fortress walls and request some ATP to aid in breaching said walls. Thus, pathogen-based pore-forming toxins provide an excellent set of candidates for autocatalytic channel-formation. This is a key attribute used for the channel-forming toxin (alpha-hemolysin) central to the nanopore detector construction described in Ch. 14.

Receptors and Messengers (Primary and Secondary)
Cells are easy units of selection, so it's easy to imagine that other membrane constituents, than for 'basic' maintenance of chemical composition, might be developed. Thus it is not surprising to find that cells communicate, and evidently with selection advantage if they can communicate with each other cleverly (about availability of resources, for

example, such as with quorum sensing). The molecular mechanism of cell-cell communication involves what are referred to as messenger molecules. Messenger molecules typically involve certain short proteins, referred to as peptides, that specifically bind (the primary messenger) to cell-surface receptors. Once bound, the cell-surface receptor relays the 'message' into the cell interior, often by activating enzymatic activity (with substrate the secondary messenger) or by means of releasing or binding a short protein or peptide on the cell interior (a more complex secondary messenger). The secondary messenger, in turn, can lead to cell growth, division, or death. The secondary messenger can also trigger production of particular molecules (of great interest from a biotechnology perspective). One of the most common cell-surface receptors is the G-protein coupled receptor (GPCR), to which about 40% of medications are targeted. (The G-protein is a heterotrimeric protein and occur in the context of the genome structure of operons).

Not all cell receptors need be on the surface, especially those that modulate gene expression. In these cases, the transport of the messenger to the receptor in the cell interior can happen by a variety of means, and once bound to receptor, an activated receptor message or action (binding or enzymatic, for example) then ensues. The typical peptide hormone or neurotransmitter 'first messenger' is hydrophilic, so cannot cross the cell membrane barrier. The steroid hormones are an example of first messenger that can cross the membrane barrier and directly interact with an interior receptor.

The step going from primary messenger to secondary messenger can be one-to-many, and can lead to further downstream catalytic cascades in the cell interior. Thus, there are multiple stages at which signaling amplification can occur.

Not all primary messengers need be protein or peptide based. Perhaps the most famous non-protein system is the endocannabinoid system, which is comprised at two endocannabinoid receptors (CB1 and CB2) and two GPCRs (so far), and a variety of endogenous arachidonate-based lipids (ancient lipids in the plant kingdom) that interact with these receptors, known as endocannabinoids. Famous endocannabinoids from exogenous sources include THC from the cannabis plant.

Co-factors and Prosthetic Groups

Sometimes protein function requires hetero-atom configurations, where additional elements are incorporated, such as simple metals (co-factors) or small organic molecules (co-enzymes if protein and enzyme). Together with PTM, this allows diverse function not easily obtained by pure amino acid sequence otherwise, such as: photo-activation, charge transfer, and electrostatic stabilization, to name a few.

If the added small organic molecule is meant to bind to the protein at all times to achieve a stabilized/modified 3D structure, then it is known as a prosthetic group. The ribosome, a nucleoprotein, is an excellent example of the latter, where more archaic organisms have ribosomes predominantly nucleic acid based, with a small number of protein prosthetics. While later organisms, eukaryotes and multicellular, for example, exhibit more and greater sophistication of prosthetics, such that the ribosome becomes a more protein-based assemblage (although the active sites remain in the ancient nucleic acid portions).

Structure-Function Paradigm

The classic structure-function paradigm has been a critical part of the understanding and exploration of protein function for over 100 years (since the advent of x-ray crystallography at the start of thc 20th Century). The gold-standard for structure understanding has been crystallography-based for all of this time, and has been critical to understanding proteins. There are problems with the crystallography-based approaches, however, first you have to have a crystal, which means you have to start with something solution-soluble. Furthermore, the crystal conformer-structure can't be too different from native conformation or the effort to identify structure is wasted. This is more complicated than it sounds, consider first the non-solution soluble (membrane-bound) proteins, their structure requires locking with a molecule to draw it into a soluble state (a detergent, essentially), such that the detergent-bound form is what is crystallized. If that form is carefully selected to be close to native then all is good, but this can take several attempts, each requiring many months. Consider also, that the solution soluble conformer structure works well as a clearly defined object, without significant variation, only for proteins consisting of roughly 200 amino acids or less, thus the larger molecules must be broken and crystallized in parts. This latter complication does not bode well for understanding multiple conformations of the (allosteric) molecule at different instances in time when we can't even get a clear picture of the entire molecule at a single instant in time. Another complication of the protein structure approach is that identifying any structure by this process is expensive and time consuming, and simply doesn't scale to the rapidly growing task given the rate of new protein discovery. If may be that a multi-nanoscope assayer, leveraging existing multi-well reactor sophistication (such as with ZMW sequencing), would allow scalable protein function analysis, but in doing so we would of necessity move from a structure-function paradigm to a *binding-state* function paradigm.

Changing Structure – Proteins as Dynamic entities

For many proteins, their dynamics is critical to their function. This is most notably the case for allosteric enzymes, for which binding analyte induces conformational change, which changes the enzymes function, and possible further binding (if it has multiple binding sites, such as for the quaternary structured hemoglobin or streptavidin molecules). Often it's the natural dynamics of the protein that is modulated by co-factors, or other molecules. Given this critical new information, and the need for characterizing more complex molecules and mixtures, such as complex glycoprotein mixtures (well beyond the scope of a crystallography effort), we see the need for a radically new approach.

State-Function Paradigm (via Nanoscope)

The Nanoscope (described extensively in [505,507]), may provide a key way forward in the 'post-genomic' era, where scalable inexpensive procedures are needed to assay complex mixtures. One of the key benefits may be realized by a shift from the protein structure-function paradigm, where we know the structure and infer the function to a "binding state"-function paradigm, where we know the binding state and infer the function (where binding state is knowledge that can be obtained via a scalable process).

B.8.2 Optimized proteins

There is a recent and bizarre connection from mathematical communications theory [2,3,15] that should be mentioned, so first a mathematical interlude.

In mathematics papers on information propagation [1-4] we find that maximal information can be propagated (with unitary property maintained) building from the standard 8-dimensional octonion, but with two more dimensions via chiral extensions (into the sedenions and trigintaduonions). This is, thus, describing a 10-dimensional propagation in a 32-dim. space, meaning that propagation in 22 of the 32-dim. does not occur, giving rise to 22 emergent parameters instead.

Thus information propagates with two chiralities, thus 2x2=4 'types', and does so in a context where 22 parameters are emergent in describing that propagation. Any highly optimized (proven effective via usage) information encoding scheme will, arguably, begin to take on the aforementioned structure for maximal transmission of information. For the connections to Physics, see the papers [1-4]. Recall that the ancient written languages all had 22 letters, in analysis of thought (sequences) there are 22 standard Jungian archetypes, and in Tarot readings, seen as a story generation mechanism, there is a deck with 22 major arcana (and 52 minor arcana). Words and 'stories' are derived from sequences of these 22 letters, archetypes, or arcana. Is there an optimal word 'size'? In Babylonian and gematria the answer is yes, where size is determined by a letter weighting scheme with optimal weighting determined to be ~137 (this same optimal is proven to exist mathematically [1-4] and in Physics, where in inverted form, it is known as the famous alpha = 1/137 parameter).

Now, let's return to proteins seen as a sequence of (22) amino acids, an encoding scheme that has been selected after billions of years of optimization (with a massively parallel Avogadro number, 10^{23}, of parallel chemical processes in every water drop). Arguably the fact that it would allow maximal information transmission with 22 letters (amino acids) in making up its words could be a coincidence….. But the coincidence continues: there are 4 (chiral) modes of information transmission, or in the case of proteins, there are thus four protein families thereby indicated – which is the case [511]. Also, the obvious 'tokenization' of protein coding in terms of exons has remained a mystery for decades. Originally it was thought that proteins would have some structural or functional modularization according to their exon 'chunks', but it turns out this is often not the case – the size and composition of exons appears to obey rules, but we don't know those rules. In the context of optimal information transmission we may finally have an answer, it may be that there is an optimization on 'protein-gematria' score (to be 137 in some amino acid weighting scheme), that is evaluated on the exons as regular gematria evaluates on individual words. The (unknown) protein gematria scoring is probably closely related to hydrophobicity 'scoring' (that is sometimes done), and probably relates to an optimality that gives rise to a globular protein shape with mostly neutral charge and balanced hydrophobicity (hydrophobicity more on interior of protein and hydrophobic residues more prevalent on the surface of the protein).

Encoding

So, suppose you're a clever organism that's settled on a DNA-based information encoding scheme (with four letters and a wobbly phosphodiester bond formation to polymerize). As suggested by the above, to have more and more functionalization of a protein 'helper', a new amino-acid (AA) based polymer space would be explored, one that might eventually settle on an encoding with a 22-element (AA) alphabet (where now the polymerization linker is the rigid peptide bond – allowing easier reproducibility and stability of complex structures). At this point, we would then need a mapping from a 4-base encoding polymer to a 22-base functionalizable polymer. Two nucleic acid bases could encode 4x4=16 possible AA's, not enough, so we must go to a 3 nucleic acid encoding scheme (the codon) with its unavoidable degeneracy of 64 mapping to 22. One of the codon's, the stop, is exclusively coding for "stop' an action on the translation, not an encoding mapping. Organisms only using the core 20 AA's have three stop codon types (in terms of DNA alphabet they are: TAA, TAG, TGA), thus the mapping is from 61 codons into the 20 AA's (shown in Ch. 4). More ancient (single celled) organism can have the full 22 AA's set, with the same exact encoding scheme for the 20 AA's, but now two of the three 'stop' codons are switched to encoding for the 21st and 22nd AA's.

B.8.3 Glycoproteins and PTMs

Determining the glycosylation profile of antibodies, and Fc glycosylation in particular, is critical to understanding antibody efficacy and blood circulation half-life, so the nanopore platform and the same signal processing methods for understanding NTD transducers can be directly applied to profiling antibody glycosylation blockade signals where the antibody is treated as an NTD transducer in and of itself. Direct antibody profiling would likely only work for part of the glycosylation (or glycation) profile, however, since the Fab N-terminus neutral glycosylation and glycations would probably still need to be assayed by use of antibody intermediates (as with the standard HbA1c test).

Antibodies are the secreted form of an associated B-cell receptor, where the difference between receptor and secreted forms is in the C-terminus of the heavy chain region. Fig. B.21 shows the standard antibody schematic and a typical antibody N-glycosylation (exact example for equine IGHD

[512]). The main non-enzymatic glycations occur spontaneously at lysines ('K') in proteins in the blood stream upon exposure to glucose via the reversible Maillard reaction to form a Schiff Base (cross-linking and further reactions, however, are irreversible and associated with the aging process).

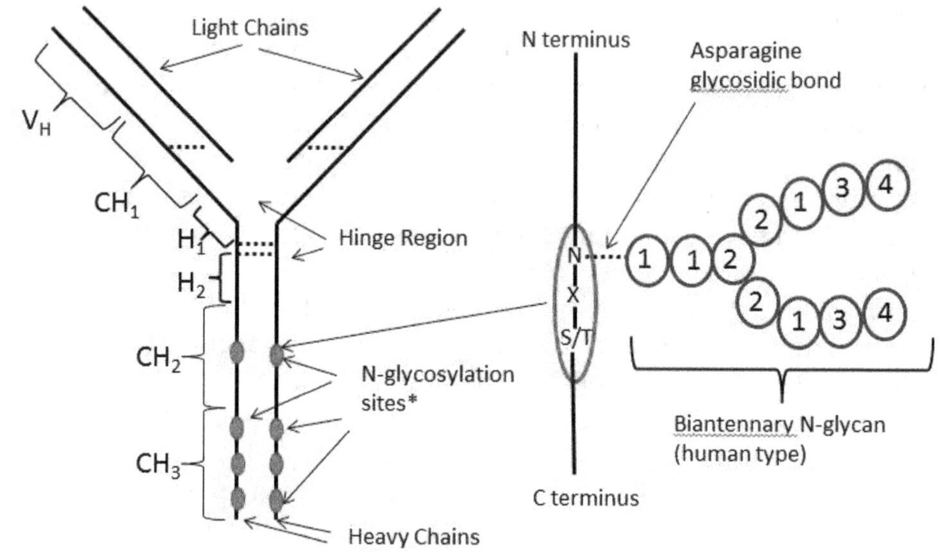

Figure B.21. The standard antibody schematic and a typical (human) antibody N-glycosylation.

The C-termini and Fc glycosylations of an antibody's heavy chain, especially for IgG, is a highly selected construct that appears to be what is recognized by immune receptors, and is evidently what is recognized as distinct channel modulator signals in the case of the NTD (mAb channel blockade signals are shown in Ch. 14). Using NTD we can co-opt the opsonization receptor-binding role of the Fc glycosylations (and mAB glycations and glycosylations in general), and C-terminus region, to be a channel modulating role. This may also permit a new manner of study of the critical opsonization role of certain classes of antibodies (and possibly differentiate the classes in more refined ways) by use of the nanopore detector platform. The channel may provide a means to directly measure and characterize antibody Fc glycosylations, a critical quality control needed in antibody therapeutics to have correct human-type glycosylation profiles in order to not (prematurely) evoke an immunogenic response. Some of the antibody blockade signals are deep blockades like the nine-base-pair DNA hairpin blockades. Working with these very clear signals, that have stable modulating blockades that go for hours, experiments were done with addition of small amounts of antigen [513]. One of the largest human-type Fc glycosylations, G2-A2+Bi+F, has a molecular weight about 2.7 kDa, which is about half the molecular weight of the eight base-pair hairpin, so ballpark similar size for similar blockade. It would appear that a variety of C-terminus modifications and Fc glycosylations would be appropriately sized and charged for channel capture, and this is what is seen in practice (see Ch. 14 for results).

B.8.4 conformation and individual-aspect – isomer resolution role for the Nanoscope

Every conformation and functionalization, thus interactivity, is explored in the 3-D protein space. There are an estimated 10^4 protein structure families [511], together with an average number of structures per family at 10^2 (this is in a power-law distribution so have extremes of only a few in a family to thousands). There are thus and estimated 10^6 uniquely functionalized protein structures.

Another family of protein-based conformations/functionalizations is that provided by the immune system's generation of immunoproteins. The variety of immunoproteins, the "immune repertoire" [514,515], appears to be a measure of health and overall age. In a healthy young adult, the immune repertoire is estimated to be as many as 10^6 uniquely functionalized protein structures (in agreement with the general protein structure-space count mentioned above). In older and sickly patients the immune repertoire can be as little as 10^3 to 10^4 structural variants (thus less effective immune response with age or disease). Having the means to effectively profile the immune repertoire of a patient may provide a strong method for overall immune system health evaluation in the tests of the future.

Resolving the up to 10^6 structural variants appears to be impossible, but the problem can be reduced by understanding the generation of these biomolecules. In other words, the information on the mix of 10^6 structural variants is related to the genomic-level information that precedes. The genome information is only part of the information used in identifying the protein families, but assume this and other information allows the 10^6 structural variants to be split back into their 10^4 protein structure families (of roughly 10^2 unique structures each). In this case the problem is reduced (with some initial genome-level family characterization, etc.) to separating roughly 10^2 unique structures. Unfortunately, albeit unique, the

10^2 unique structures are part of the same family and thus very similar (insofar as active-site conformation). The problem of resolving a family of 100-200 very similar biomolecules (even isomers) is not just encountered here in the context of immunoproteins, but also in the attempts to assay the different cannamimetic compounds present in medicinal marijuana (an assay on small, 21 carbon, biomolecules). The main tool for doing this does not exist, but there is possible applications of the Nanoscope for this purpose.

B.8.5 Gene Circuits and the Nanoscope biosystem 'voltmeter'
Biosystem circuits come in a variety of forms: (i) metabolic cycles; (ii) info/metabolic such as cycles mRNA→protein (where protein state/dynamics enumerates functionality); and (iii) info cycles: DNA → mRNA (typically involving DdRp, DdDp, RdRp, or RdDp).

The system biologist is currently lacking a general-use method for a biosystem circuit or gene circuit 'voltmeter' or biosystem algorithm 'print statement'. What is needed is a non-destructive, carrier non-modifying, means of testing 'live' biological systems at the single-molecule level. A method using the nanopore transduction detector (NTD) is demonstrated for single-molecule characterization in some situations, so may provide what is lacking. An important aspect of this approach is that use can bc made of inexpensive antibody, protein, aptamer, duplex nucleic acid, or nucleic acid annealing molecules (for miRNA and viral monitoring) that have *specific* binding to the system component of interest. The NTD transducer's specific binding can also be designed to have low affinity binding as needed, such that there can be a 'catch and release' on low copy-number molecular components, such that there is not a disruption to the molecular system under study. NTD transducers are typically constructed by linking a binding moiety of interest to a nanopore current modulator, where the modulator is designed to be electrophoretically drawn to the channel and partly captured, with its captured end distinctively modulating the flow of ions through the channel. Using inexpensive (commoditized) biomolecular components, such as DNA hairpins, this allows for an easily constructed, versatile, platform for biosensing. High specificity high affinity binding also allows a very versatile platform for assaying at the single molecule level, even down to the single isoform level, including molecular substructure profiling, such as glycosylation profiling in antibodies. An inexpensive commoditized pathway for constructing nanopore transducers is demonstrated. Nanopore transduction detector based reporter/event-transducer molecules may serve as a means to perform multicomponent mRNA-miRNA-protein and protein-protein systems analysis in general settings.

A growing number of questions facing molecular and medicinal biology experts are systems biology questions, where the complex interaction of genes, mRNAs, proteins, miRNAs, and various metabolites is described at the 'system level'. System level problems are often described in terms of 'gene circuits' or 'metabolic algorithms'. These comparisons to system descriptions in electrical engineering and computer science offer some insights due to actual parallels, and some misleading comparisons due to oversimplification in comparison to actual biological systems.

A reductionist analysis of a biological system, not surprisingly, reveals that the sum is greater than its parts. But this is actually found to be the case in electrical circuits as well, where emergent properties, especially emergent noise and communications properties, are often found in circuits with feedback. Even simple physical systems involving just three bodies in classical orbital dynamics gives rise to chaotic behavior, which was not expected in early physics, where the sum was originally NOT thought to be greater than its parts. Iterative dynamical systems in general are found to exhibit chaotic behavior and emergent constructs such as strange attractors and limit cycles. Systems with feedback, thus, can do surprising things, and biological systems definitely have done some surprising things ranging from living systems in their amazing variety to complex phenomena such as intelligence, language, and consciousness.

The nanopore transduction detector (NTD) method is typically based on a single protein-channel biosensor implemented on a lipid bilayer (synthetic cell membrane), but it could also be implemented as a live cell assay by using the original patch clamp protocol for measuring current through a channel on a live cell [516] (the invention of the patch clamp amplifier resulted in the Nobel Prize for Medicine or Physiology in 1991 [517]). In order for the NTD 'voltmeter' operating on the biological system to work, cell-based or not, the normal operational buffer of the NTD must accommodate a change to the physiological or cellular buffer environment of the biological system of interest, and, if cell-based, the 'carrier signal' that is the basis of the analysis can no longer be channel-current based, but channel-noise based with use of laser modulations for noise state excitation. Work with robust NTD operation with a variety of buffer pH and in the presence of high concentrations of interference agents reveals that operational stability with a wide range of buffers has been achieved [518]. Laser modulations have also been introduced to improve the NTD mechanism to have more general applicability [519], and for purposes of establishing an improved 'stochastic carrier wave' molecular state tracking capability [505], so many of the complications with returning to the single-cell application are mostly solved. What remains to be resolved for general applicability of the NTD system analyzer method, for both *in vitro* and an possible *in vivo* studies, is a standardized method for NTD transducer construction and operation (see [520]). An inexpensive method for a NTD-based biological system 'voltmeter' is thus possible for both *in vivo* and *in vitro* applications.

Electrical/Biological circuits: the biosystem extra element theorem (BEET)

A reductionist analysis of electrical circuits involves a reduction to circuit elements that have linear responses. In this regard biology only compares weakly, as the components of a biological circuit are generally non-linear over much of their operational range. Even so, for some biological system settings sufficiently small perturbations in the biological components can often be made such that they provide a linear system response. Given the complexity of the biological feedback systems, however, this might seem to be small progress. It is very significant, however, given the existence of a sophisticated method from advanced circuit design and analysis that is applicable for linear response systems known as the 'extra element theorem' [521]. It interesting to note that this important circuit method from electrical system theory has not been imported into biological system discussions given its likely significant role in molecular evolutionary theory. The extra element theorem from electrical circuit theory allows simpler circuits, that are more easily understood, to have new components added (the 'extra' element), and if the new component happens to create a feedback loop, then the complexity of the feedback loop analysis can be much more easily evaluated and understood directly by way of the extra element theorem. In practice, very complex electrical amplifier circuits can be built-up and analyzed in this way, by repeated use of the extra element theorem. This offers the means to have a reductionist analysis while capturing the growing complexity of holistic irreducible systems. For a biological variant of the extra element theorem a patchwork of linear response regimes could be used in understanding a particular biological system.

The 'messengers' in biological and electrical systems differ greatly in many respects, which can make some gene circuit intuition entirely misguided. The carriers in an electrical circuit, for example, are remarkably simple by comparison with biological system signal carriers. Electrical charge moves through wires like a fluid. Granted, the electrical charge moves at a sizable fraction of the speed of light, but it is so like a fluid flow that some current flow discussions are basically plumbing discussions, where the description of the current flow is often compared to flow of water through pipes where pipe narrowness is akin to resistance, etc. The flow/interaction topology of electrical current is also self-evident in the connectivity that can be seen in the wiring of the circuit diagram. If the biological system is too interconnected in this comparison this is often where the analogy is shifted to discussions of a gene system *algorithm*. The electrical messengers, or charge carriers, are also vastly simpler than the biological system messengers. Electrical current carriers are of only one type (electrons), and don't have attractive self-interaction molecular carriers (as with dimerization … unless you are talking superconductivity), and don't have internal state (in the sense of the circuit model) like with biological secondary messengers. Biological system messengers, on the other hand, come in a huge variety, operate at the single molecule level, and depending on perspective, everything in the biological system might be considered a system messenger in a massive, living, autocatalytic cascade. The biological system carriers or messengers are also much fewer in number compared to their electrical counterparts. This actually makes things more complicated. In electronics having small currents is modelled as a noise source, where once the discreteness of the charge carriers begins to be discernible this puts one in the realm of stochastic 'shot' noise. In the biological comparison this stochastic underpinning, if significant, again favors a shift to the 'algorithm' analogy instead of the circuit analogy. To further complicate matters, the biological carriers of the system interactions interact with each other, and typically have internal states (e.g., proteins and riboswitches often have conformational states), so the picture of the carriers for biology introduces vastly greater complexity and interaction interconnectivity.

In electrical circuit analysis a good voltmeter is something that will not significantly 'load' or alter the circuit while measuring a particular component's voltage drop. Likewise, in analyzing a computer program, or resolving a runtime error (the closest analogy to analyzing a 'live' biological algorithm), one of the best tools available is to simply introduce a 'print statement' to track any internal state behavior of interest in the program. This is where the weakness of the circuit or algorithm analogy in biological systems is most profound. The system biologist doesn't have a gene circuit voltmeter or gene system algorithm print statement. Some of the closest biochemistry methods to offer such capabilities are fluorescence based, and in certain specialized applications remarkable results have been obtained along these lines, but they typically involve the introduction of constructs with a great deal of effort that won't scale well to the vast number of biological systems that need to be studied in the post-genomic era. What is needed is a non-destructive, carrier non-modifying, means of testing 'live' biological systems, possibly in their native cellular environment.

Validation of miRNA's and miRNA binding sites using a nanopore transduction nanoscope

The discovery of the RNA interference (RNAi) immune response and translational regulation mechanism has led to an explosion in the number of identified microRNAs (miRNAs) and their mRNA binding sites. An understanding of miRNAs and their binding sites, typically in the 3' untranslated region (3' UTRs) of mRNAs, is helping to explain a wide range of complex phenomena, ranging from latency control by viruses during infection (such as with HIV) [522], to complex regulation in system syndromes such as in diabetes and in the effects of aging [523], to the general trans-regulation of mRNAs at the translational level (complementing transcription factor and promoter cis-regulation at the transcriptional level) [524]. The examination of miRNAs, and especially miRNA binding sites, is confounded by the small size of the miRNAs, however: 21-25 nucleotides in length for typical mature miRNAs, and only 7-8 base ssRNA seed regions in the guide-strand RNA incorporated into the RNAi's RISC complex for actual binding/repression to complementary 7-8 base sequence in the 3'UTR region of the target mRNA [525]. For the latter case of verification for miRNA/RISC derived sequence binding with a 7 base sequence in a mRNA's 3'UTR there is further complication given possible posttranscriptional modifications, such as via inosine substitution for adenosine due to adenosine deaminases with inosine recognition as guanine in terms of base-pairing that can alter the actual target sequence of the miRNA/RISC binding [526]. This is in addition to the obvious complication of identifying the presence of RNA annealing when the annealing only involves 7 bases of RNA.

Preliminary work with NTD-based detection on short DNA annealing suggests a possible means to examine the miRNA/RISC binding to target

3'UTR region with or without the RISC complexes argonaute proteins intact, where results are expected to improve even more upon refinement using locked nucleic acid transducer/reporter probes (see Discussion). NTD based detection of DNA annealing has been demonstrated on DNA sequences as short as 5 bases [527], and in the presence of a variety of interference agents and chaotropes [518]. NTD based detection has also been demonstrated in a variety of buffer conditions so could be established in a buffer conducive to the RISC complex remaining intact and where the annealing to 3'UTR complement sequence occurs with the binding strength found *in vivo*. NTD detection can also operate on small volumes since it makes use of a *single* protein channel interaction, thereby inherently operating at the single-molecule interaction level. NTD detection can, thus, identify single-molecule binding events in a non-destructive manner that may be conducive to the 'live' characterization of many critical, transient, interactions.

For biosensing or bioassays applications in general, not all miRNA or miRNA binding site analyses need be in cellular or physiological buffer either. In a 'destructive setting' more forceful miRNA validation assays, and analysis of annealing-based events, can be pursued by use of chaotropes such as urea. Clearer identification of collective binding events, such as for highly complementary annealing interactions, is found to occur upon introduction of chaotropes that eliminate non-specific DNA interactions, and many 'simple' binding interactions, not involving collective interactions of many components as with annealing [528].

Single-molecule protein assays and protocols – another role for the Nanoscope

Scalable Protein Assays are sought. At present there is a non-scalable problem when protein described according to the standard structure-function paradigm, since structure cannot be sufficiently well-determined via scalable structure resolution methods (the gold-standard for structure determination, x-ray crystallography, is not scalable). Structure determines the function, but we can have information that maps out that function without determining the structure. If that function information, or information gathering process, is scalable, then we finally can have a scalable approach to protein studies.

Note that if the 'protein' is sufficiently small, too small for a fold structure like that found in lengthier AA sequences, then it is referred to as a peptide, and peptides are used as secondary messengers, among other things. Also in this group, of small critically functionalized molecules (such as the peptides), are other small signaling molecules, often derived from the secretory pathways of the cell (ER and Golgi PTM, if protein or peptide, alpha acids, beta acids, cannabinoids, terpenes, pyrethrins). How to assay the presence of such small molecules, especially when they are very similar, if not isomers, of each other? We know we can preferentially bind (and signal this binding) molecules only differing by isomer configuration. A universal platform to take a specific binder (antibody or aptamer) for a specific target and leverage it into a biosensor for that target is what is needed. So whether mapping binding function on larger proteins or specifically binding molecular features on smaller or similar molecules (to collectively differentiate isomers, for example), a Nanoscope approach could be used to solve the problem.

Biosensing by way of specific binders, one for each individual molecule to track, cannot scale well when the number of specific targets grows too large (beyond the 100 or so specific signals that can be resolved in a given experimental setup). So, specific binders are needed that effectively bind onto less unique sub-structures or folds of the molecules. A collection of these less-specific binders then profiles what is present and may be able to detect on 100 to 1000s of molecular targets with only 10 or so biosensing binders. So scalable protein analysis appears to be within reach using the Nanoscope platform.

Appendix C. Narrative world

Narrative world provides a rich tapestry of possibilities where multiverses can exist alongside magical elements and the concept of monadic travel. The idea of multiverses suggests that there are countless parallel realities. This concept has been popularized in modern science fiction and fantasy literature, where characters often traverse different dimensions or realities. Magic can be given in a number of different forms within these narratives. It can also symbolize the unseen forces that connect different realms of existence. Monadic travel refers to the philosophical idea derived from Leibniz, where each individual (or monad) is a unique, self-contained entity that reflects the universe from its own perspective. In a narrative context, this can imply that each character's journey through different realities or experiences can effect a form of travel [529].

There are many questions raised by these issues, discussed by others, here are a few:
How to spiritually shift to the "best of all worlds"? Does it happen unconsciously amongst a group of conscious beings (normally)? Does consciousness differ from mere computation and action in that it 'selects' alternate realities? If consciousness is implicit selection, can the mind be wielded by consciousness to effect explicit selection? Does Dark matter provide the Many-Worlds classical scaffolding of a "measurement apparatus"? where the traversal of these many worlds involves use of the common Dark matter 'stage'.

Emanator Theory allows multiverses:
Our 'reality' is a surface phenomenon riding on a trigintaduonion product. There are alternate realities also 'riding' in the trigintaduonion product, that can interact, thus have many worlds linkage.

The Tipler omega point
The Tipler Omega Point is a theoretical concept proposed by physicist Frank J. Tipler in his 1980 book "The Physics of Immortality" [530]. The Omega Point is a hypothetical future state of the universe in which all consciousness is preserved and all information is stored in a highly advanced, technologically sophisticated manner.

The Omega Point is described as a state of maximum complexity and information processing, where the universe reaches a point of infinite knowledge and consciousness. Tipler posits that as the universe evolves, it will eventually undergo a process of "recycling" where all information and consciousness are preserved. Tipler's model is rooted in the laws of physics, particularly thermodynamics and cosmology. He argues that the universe will ultimately collapse in a "Big Crunch," leading to a singularity where all information is concentrated. At this point, he suggests that advanced civilizations (or emergent godhood) could manipulate the laws of physics to achieve immortality and preserve consciousness. Further discussion of this hypothesis is in [531].

References

[1] Winters-Hilt, S. (2023). Emanation, Emergence, and Eucatastrophe. (Physics Series: "Physics from Maximal Information Emanation" Book 7.)
[2] Winters-Hilt, S. (2015). Feynman-Cayley Path Integrals select Chiral Bi-Sedenions with 10-dimensional space-time propagation. Advanced Studies in Theoretical Physics, Vol. 9, 2015, no. 14, 667 – 683. dx.doi.org/10.12988/astp.2015.5881.
[3] Winters-Hilt, S. Unified propagator theory and a non-experimental derivation for the fine-structure constant. Advanced Studies in Theoretical Physics, Vol. 12, 2018, no. 5, 243-255. https://doi.org/10.12988/astp.2018.8626.
[4] Winters-Hilt, S. The 22 letters of reality: chiral bisedenion properties for maximal information propagation. Advanced Studies in Theoretical Physics, Vol. 12, 2018, no. 7, 301-318. https://doi.org/10.12988/astp.2018.8832.
[5] Winters-Hilt, S. Theory of Trigintaduonion Emanation and Origins of α and π. Researchgate 05/24/20.
[6] Winters-Hilt, S. Chiral Trigintaduonion Emanation Leads to the Standard Model of Particle Physics and to Quantum Matter. Advanced Studies in Theoretical Physics, Vol. 16, 2022, no. 3, 83-113.
[7] Winters-Hilt, S. Fiat Numero: Trigintaduonion Emanation Theory and its Relation to the Fine-Structure Constant α, the Feigenbaum Constant $C\infty$, and π. Advanced Studies in Theoretical Physics Vol. 15, 2021, no. 2, 71 - 98 HIKARI Ltd, www.m-hikari.com https://doi.org/10.12988/astp.2021.91517.
[8] Winters-Hilt, S. Meromorphic precipitation of quantum matter with dimensionful action. May 2021. DOI:10.13140/RG.2.2.32294.24640.
[9] Winters-Hilt, S. Emanator Theory using split octonions is Manifestly Lorentz Invariant and reveals why the fundamental constant $\hbar$ should be so small. Advanced Studies in Theoretical Physics, 2023.
[10] Winters-Hilt S. Emanator Theory is shown to be an optimal Martingale process at the fractal edge of chaos, where the Gravitational constant G is hypothesized to be a multiscale fractal coupling parameter. Advanced Studies in Theoretical Physics, 2023.
[11] Winters-Hilt, S. Informatics and Machine Learning. Wiley Publishing. 9781119716747, Sept. 2021.
[12] Eiseley, L. (1957). *The immense journey*. Random House.

[13] Jacobsen, T. (1939). The Sumerian King List. Assyriological Studies, 11, 1-34.
[14] Vaswani, A., Shazeer, N., Parmar, N., Uszkoreit, J., Jones, L., Gomez, A. A., Kaiser, Ł., & Polosukhin, I. (2017). *Attention is all you need*. Advances in Neural Information Processing Systems, 30. Retrieved from https://arxiv.org/abs/1706.03762
[15] Kaplan, J., McCandlish, S. H., Henighan, T., Brown, T. B., Chess, B., Child, R., ... & Amodei, D. (2020). *Scaling laws for neural language models*. OpenAI. Retrieved from https://arxiv.org/abs/2001.08361
[16] Balmer, J. J. (1885). "Notiz über die Spectrallinien des Wasserstoffs" [Note on the spectral lines of hydrogen]. Annalen der Physik und Chemie. 3rd series (in German). 25: 80–87.
[17] Sommerfeld, A., *Atombau und Spektrallinien* (Friedrich Vieweg und Sohn, Braunschweig, 1919).
[18] Wolfgang Pauli – Nobel Lecture. NobelPrize.org. Nobel Media AB 2021. Tue. 2 Mar 2021.
https://www.nobelprize.org/prizes/physics/1945/pauli/lecture/
[19] Feynman, R.P. QED: The Strange Theory of Light and Matter. Princeton University Press. p. 129. (1985) ISBN 978-0-691-08388-9.
[20] Mandelbrot, B. B. (1983). *The Fractal Geometry of Nature*. W.H. Freeman and Company.
[21] Falconer, K. (2003). *Fractal Geometry: Mathematical Foundations and Applications*. Wiley.
[22] Barnsley, M. F. (1993). *Fractals Everywhere*. Academic Press.
[23] Peitgen, H.-O., Jürgens, H., & Saupe, D. (1992). *Chaos and Fractals: New Frontiers in Science*. Springer.
[24] Hutchinson, J. E. (1981). "Fractals and self-similarity." *Indiana University Mathematics Journal*, 30(5), 713-747.
[25] Barnsley, M. F. (1988). "Fractal Image Compression." *Proceedings of the IEEE*, 76(6), 836–845.
[26] S. A. Kauffman (1993). *The Origins of Order: Self-Organization and Selection in Evolution*. Oxford University Press.
[27] G. A. Edgar (1990). *Measure, Topology, and Fractal Geometry*. Springer.
[28] Edgar, G. A. (2008). Self-similarity and the structure of fractals. *Mathematics Magazine*, 80(1), 15–27.
[29] Sornette, D. (2006). *Critical Phenomena in Natural Sciences: Chaos, Fractals, Self-Organization and Disorder: Concepts and Tools*. Springer.

[30] Bak, P., Tang, C., & Wiesenfeld, K. (1988). "Self-organized criticality: An explanation of 1/f noise." *Physical Review Letters*, 59(4), 381-384.
[31] Feder, J. (1988). *Fractals*. Plenum Press.
[32] J. P. Sethna, K. A. Dahmen, & C. R. Myers (2001). "Crackling noise." *Nature*, 410(6825), 242-250.
[33] G. A. Edgar (2008). "Fractal Geometry: Mathematical Foundations and Applications." *Mathematics Magazine*, 81(5), 353-366.
[34] Winters-Hilt, S. (2024). Thermal & Statistical Mechanics, and Black Hole Thermodynamics. (Physics Series: "Physics from Maximal Information Emanation" Book 6.)
[35] Winters-Hilt, S. (2024). Quantum Field Theory and the Standard Model. (Physics Series: "Physics from Maximal Information Emanation" Book 5.)
[36] S. M. de Jong, et al. (2009). "Fractal analysis of the spatial distribution of galaxies." *Astronomy & Astrophysics*, 507(2), 1007-1015.
[37] Winters-Hilt, S. (2023). The Dynamics of Manifolds. (Physics Series: "Physics from Maximal Information Emanation" Book 3.)
[38] D. H. Weinberg, et al. (2013). "The Luminous and Dark…
[39] Peebles, P. J. E. (1980). *The Large-Scale Structure of the Universe*. Princeton University Press.
[40] J. A. Peacock (1999). *Cosmological Physics*. Cambridge University Press.
[41] Feigenbaum, M. J. (1976) "Universality in complex discrete dynamics", Los Alamos Theoretical Division Annual Report 1975-1976.
[42] McMullen, Curtis T. 2000. The Mandelbrot set is universal. In The Mandelbrot Set, Theme and Variations, ed. T. Lei, 1–18. Cambridge U.K.: Cambridge Univ. Press. Revised 2007.
[43] Rao, V. S., & Kuo, C. C. J. (2005). "Unit-norm propagation for sparse signal recovery." *IEEE Transactions on Signal Processing*, 53(12), 4550-4560.
[44] Liu, J., & Wang, L. (2015). "Unit-norm optimization for sparse representation." *Journal of Machine Learning Research*, 16, 1-30.
[45] Caves, C.M., C.A., Fuchs, R. Schack. Unknown quantum states: The Quantum de Finetti Representation. J. Math. Phys. 43, 4537 (2002).
[46] Gunaydin, M. and F. Gursey. Quark structure and the octonions. J. Math. Phys., 14, 1973.
[47] Hurwitz, A. (1923), "Über die Komposition der quadratischen Formen", Math. Ann., 88 (1–2): 1–
25, doi:10.1007/bf01448439, S2CID 122147399.

[48] Dixon, G.. Division algebras: octonions, quaternions, complex numbers and the algebraic design of physics. Kluwer Academic Publishers, 1994.
[49] Furey,C.; Towards a unifed theory of ideals; arXiv:1002.1497v5 [hep-th] 25 May 2018.
[50] Harmon, P.M., Editor, Scientific Letters and Papers of James Clerk Maxwell, Vol. II, Cambridge University Press, 1995, pp. 570–576.
[51] Hankins, T.L., Sir William Rowan Hamilton, Johns Hopkins Press, 1980, pp. 316–319.
[52] Dyson, F.J., Feynman's proof of the Maxwell equations, Am. J. Phys., 1990, 58, 209-211.
[53] Dyson, F.J., Feynman at Cornell, Phys. Today, 1989, 42 (2), 32-38.
[54] Hughes, R.J., On Feynman's proof of the Maxwell equations, Am. J. Phys., 1991, v. 60(4), 301.
[55] Silagadze, Z.K., Feynman's derivation of Maxwell equations and extra dimensions. Annales de la Fondation Louis de Broglie, 2002, v. 27, no.2, 241.
[56] Manogue, C.A. and J. Schay, Finite Lorentz Transformations, Automorphisms, and Division Algebras, Hep-th/9302044.
[57] Manogue, C.A. and A. Sudbery, General solutions of covariant superstring equations of motion, Phys. Rev. 40 (1989) 4073.
[58] Cederwall, M., Octonionic particles and the S7 symmetry, J. Math. Phys. 33 (1992) 388.
[59] Gogberashvili, M., Octonionic electrodynamics, J. Phys. A: Math. Gen. 39 7099.
[60] Chanyal, B.C., P. S. Bisht and O. P. S. Negi, Generalized Split-Octonion Electrodynamics, 2010, arXiv:1011.3922v1.
[61] Kravchenko, V., Quaternionic equation for electromagnetic fields in inhomogenous media, arXiv:math-ph/0202010.
[62] Smith, Jr., F.D. Standard Model plus Gravity from Octonion Creators and Annihilators, Quant-ph/9503009.
[63] Pushpa, P.S. Bisht, T. Li, and O. P. S. Negi, Quaternion Octonion Reformulation of Quantum Chromodynamics, Int. J. Theor. Phys., 2011, Vol 50, 2, pp 594-606.
[64] Mironov, V.L., and S. V. Mironov, Associative Space-Time Sedenions and Their Application in Relativistic Quantum Mechanics and Field Theory, Applied Mathematics, 2015, 6, 46-56.
[65] Chrisitianto, V., F. Smarandachey, A Derivation of Maxwell Equations in Quaternion Space, April, 2010 PROGRESS IN PHYSICS Volume 2.

[66] Conway, J.H. and D.A. Smith, On Quaternions and Octonions: their geometry, arithmetic, and symmetry, A K Peters, Wellesley, Massachusetts, 2003.
[67] RCHO(ST) Hypothesis, http://www.meta-logos.com/HTQG_100811.pdf , 2011.
[68] Winters-Hilt, S. (2023). Classical Mechanics and Chaos. (Physics Series: "Physics from Maximal Information Emanation" Book 1.)
[69] Winters-Hilt, S. (2023). The Dynamics of Fields, Fluids, and Gauges. (Physics Series: "Physics from Maximal Information Emanation" Book 2.)
[70] Winters-Hilt, S. (2023). The Dynamics of Manifolds. (Physics Series: "Physics from Maximal Information Emanation" Book 3.)

[71] Bak, P. (1996). *How Nature Works: The Science of Self-Organized Criticality.*
[72] Langton, C. G. (1990). "Computation at the Edge of Chaos." In *Artificial Life II*, edited by Langton, C. G.
[73] Holland, J. H. (1995). *Hidden Order: How Adaptation Builds Complexity.*
[74] Kauffman, S. A. (1993). *The Origins of Order: Self-Organization and Selection in Evolution.*
[75] Mitchell, M. (2009). *Complexity: A Guided Tour.*
[76] Sullivan, John. *Fractals and the Kabbalah: A New Perspective on the Universe*. New York: Fractal Publishing, 2010.
[77] Klein, Moshe. *The Fractal Nature of Reality: Kabbalah and Modern Science*. New York: Kabbalah Publishing, 2012.
[78] Jung, C. G. (1964). *Man and His Symbols.*
[79] Campbell, J. (1949). *The Hero with a Thousand Faces.*
[80] Montola, M., Stenros, J., & Waern, A. (2009). *Pervasive Games: Theory and Design.*
[81] Wood, J. (2003). *The Tarot: History, Symbolism, and Divination.*
[82] Winters-Hilt, S. Quantum Mechanics, Path Integrals, and Algebraic Reality. (Physics Series: "Physics from Maximal Information Emanation" Book 4.)
[83] DeepSeek-AI. DeepSeek-V3 Technical Report. arXiv:2412.19437v2 [cs.CL] 18 Feb 2025.
[84] Daya Guo et al. DeepSeek-R1: Incentivizing Reasoning Capability in LLMs via Reinforcement Learning. arXiv:2501.12948 [cs.CL]
[85] Gonzalez, G. (1998). "The Sun: A Stable Star." *Astrophysical Journal*, 496(1), 1-10.

[86] Hertzsprung, E. (1924). "The Classification of Stars." *Annual Review of Astronomy and Astrophysics*.
[87] Salaris, M., & Cassisi, S. (2005). *Evolution of Stars and Stellar Populations*. Wiley.
[88] Kaler, J. B. (2002). *The Hundred Greatest Stars*. Copernicus Books. ISBN: 978-0387954363.
[89] Hathaway, D. H. (2010). The Solar Cycle. *Living Reviews in Solar Physics*, 7(1), 1-65. https://doi.org/10.12942/lrsp-2010-1
[90] Milankovitch, M. (1941). *Kanon der Erdbestrahlung und seine Anwendung auf das Eiszeitenproblem* [Canon of Insolation and the Ice-Age Problem]. Special Publication 132, Section of Mathematical and Natural Sciences, Vol. 33. Royal Serbian Academy of Sciences. (Translated by the Israel Program for Scientific Translations, 1969).
[91] Petit, J. R., Jouzel, J., Raynaud, D., Barkov, N. I., Barnola, J.-M., Basile, I., Bender, M., Chappellaz, J., Davis, J. C., Delaygue, G., Delmotte, M., Kotlyakov, V., Legrand, M., Lipenkov, V. Y., Lorius, C., Pépin, L., Ritz, C., Saltzman, E., & Stievenard, M. (1999). Climate and Atmospheric History of the Past 420,000 Years from the Vostok Ice Core, Antarctica. *Nature*, 399(6735), 429–436. https://doi.org/10.1038/20859
[92] Rahmstorf, S. (2002). Ocean Circulation and Climate During the Past 120,000 Years. *Nature*, 419(6903), 207–214. https://doi.org/10.1038/nature01090
[93] Karlin, S., & Taylor, H. M. (1975). *A First Course in Stochastic Processes* (2nd ed.). Academic Press. ISBN: 978-0123985521.
[94] Weyl, H. (1917). Zur Gravitationstheorie [On the Theory of Gravitation]. *Annalen der Physik*, 359(18), 117–145. https://doi.org/10.1002/andp.19173591804
[95] Eddington, A. S. (1931). Preliminary Note on the Masses of the Electron, the Proton, and the Universe. *Mathematical Proceedings of the Cambridge Philosophical Society*, 27(1), 15–19. https://doi.org/10.1017/S0305004100009269
[96] Narlikar, J. V. (1977). *The Structure of the Universe*. Oxford University Press. ISBN: 978-0192176530.
[97] Cetto, A. M., de la Peña, L., & Santos, E. (1986). Dirac's Large-Number Hypothesis Revised. *Astronomy and Astrophysics*, 164, 1–5.
[98] Raychaudhuri, A. K. (1979). *Theoretical Cosmology*. Clarendon Press. ISBN: 978-0198514626.
[99] Jordon, P., The origin of stars, Der Verlag, Stuttgart, 1947.
[100] Shemi-Zadeh, V. E. Coincidence of large numbers, exact value of cosmological parameters and their analytical representation, arXiv:gr-qc/0206084, 2002.

[101] Peacock, J.A. Cosmological physics, Cambridge Univ. Press, 1999.
[102] Andreev, A.Y., and B.V. Komberg. Cosmological parameters and the large numbers of Eddington and Dirac, Astron.
Reports, 44, pp. 139-141, 2000.
[103] Ray, S., U. Mukhopadhyay, P.P. Ghosh. Large number hypothesis: A review, arXiv: 0705.1836, 2007.
[104] Valev, D. Evidence of Dirac Large Number Hypothesis. Proceedings of the Romanian Academy, Series A, Volume 20, Number 4/2019, pp. 361–368.
[105] Onsager, Lars. "Crystal statistics. I. A two-dimensional model with an order-disorder transition", *Physical Review*, Series II, 65 (3–4): 117–149, 1944.
[106] Plischke, Michael and Birger Bergersen. Equilibrium Statistical Physics. World Scientific. 1989.
[107] Chandler, David. Introduction to Modern Statistical Mechanics. Oxford University Press. 1987.
[108] Peierls, R.E. (1936). Proc. Cambridge Philos. Soc. 32:477.
[109] Wannier, G.H. (1966). Statistical Physics. New York: Wiley.
[110] Schultz, T. D., Mattis, D. C., & Lieb, E. H. (1964). Two-dimensional Ising model as a soluble problem of many fermions. *Reviews of Modern Physics*, 36(3), 856–871.
https://doi.org/10.1103/RevModPhys.36.856
[111] Landau, L. D., & Lifshitz, E. M. (1980). *Statistical Physics, Part 1* (3rd ed.). Pergamon Press. https://www.amazon.com/Statistical-Physics-Part-Revised-Enlarged/dp/0080230385
[112] Kadanoff, L. P. (1966). Scaling laws for Ising models near TcT_cTc. *Physics Physique Fizika*, 2(6), 263–272.
https://doi.org/10.1103/PhysicsPhysiqueFizika.2.263
[113] Wilson, K. G. (1975). The renormalization group: Critical phenomena and the Kondo problem. *Reviews of Modern Physics*, 47(4), 773–840. https://doi.org/10.1103/RevModPhys.47.773
[114] Rosenfeld, J. S., Rosenfeld, A., Belinkov, Y., & Shavit, N. (2019). A constructive prediction of the generalization error across scales. arXiv preprint arXiv:1909.12673. https://arxiv.org/abs/1909.12673
[115] Huang, Y., Cheng, Y., Bapna, A., Firat, O., Chen, M. X., Chen, D., Lee, H., Ngiam, J., Le, Q. V., Wu, Y., & Chen, Z. (2018). GPipe: Efficient training of giant neural networks using pipeline parallelism. arXiv preprint arXiv:1811.06965. https://arxiv.org/abs/1811.06965
[116] Shazeer, N., Cheng, Y., Parmar, N., Tran, D., Vaswani, A., Koanantakool, P., Hawkins, P., Lee, H., Hong, M., Young, C., Sepassi, R., & Hechtman, B. (2018). Mesh-TensorFlow: Deep learning for

supercomputers. arXiv preprint arXiv:1811.02084. https://arxiv.org/abs/1811.02084
[117] Child, R., Gray, S., Radford, A., & Sutskever, I. (2019). Generating long sequences with sparse transformers. arXiv preprint arXiv:1904.10509. https://arxiv.org/abs/1904.10509
[118] Gray, S., Radford, A., & Kingma, D. P. (2017). GPU kernels for block-sparse weights. OpenAI. https://cdn.openai.com/blocksparse/blocksparsepaper.pdf
[119] Krizhevsky, A., Sutskever, I., & Hinton, G. E. (2012). ImageNet classification with deep convolutional neural networks. *Advances in Neural Information Processing Systems*, 25, 1097–1105. https://papers.nips.cc/paper/4824-imagenet-classification-with-deep-convolutional-neural-networks
[120] Wang, Y.-X., Ramanan, D., & Hebert, M. (2017). Growing a brain: Fine-tuning by increasing model capacity. In *Proceedings of the IEEE Conference on Computer Vision and Pattern Recognition* (pp. 2471–2480). https://arxiv.org/abs/1907.07844
[121] Wen, W., Yan, F., & Li, H. (2019). AutoGrow: Automatic layer growing in deep convolutional networks. arXiv preprint arXiv:1906.02909. arXiv
[122] Kaplan, J., McCandlish, S. H., Henighan, T., Brown, T. B., Chess, B., Child, R., ... & Amodei, D. (2020). Scaling laws for neural language models. arXiv preprint arXiv:2001.08361.
[123] Altmann, E. G., Cristadoro, G., & Degli Esposti, M. (2012). On the origin of long-range correlations in texts. *Proceedings of the National Academy of Sciences*, 109(29), 11582–11587. PNAS
[124] Hestness, J., Narang, S., Ardalani, N., Diamos, G., Jun, H., Kianinejad, H., ... & Zhou, Y. (2017). Deep learning scaling is predictable, empirically. arXiv preprint arXiv:1712.00409. ACM Digital Library+3DBLP+3aXi+3
[125] Devlin, J., Chang, M.-W., Lee, K., & Toutanova, K. (2019). BERT: Pre-training of deep bidirectional transformers for language understanding. *Proceedings of the 2019 Conference of the North American Chapter of the Association for Computational Linguistics: Human Language Technologies*, 1, 4171–4186.
[126] Radford, A., Narasimhan, K., Salimans, T., & Sutskever, I. (2018). Improving language understanding by generative pre-training. OpenAI. OpenAI CDN
[127] Winters-Hilt, S. (2019). *Data Analytics, Bioinformatics, and Machine Learning*. Golden Tao Publishing.

[128] Miller, S. L., & Urey, H. C. (1959). Organic compound synthesis on the primitive Earth. *Science*, 130(3370), 245–251.
[129] Kauffman, S. A. (1986). Autocatalytic sets of proteins. *Journal of Theoretical Biology*, 119(1), 1–24.
[130] Szostak, J. W., Bartel, D. P., & Luisi, P. L. (2001). Synthesizing life. *Nature*, 409(6818), 387–390.
[131] Ranjan, S., & Sasselov, D. D. (2016). The role of polycyclic aromatic hydrocarbons in the origins of life. *Astrobiology*, 16(1), 1-12. https://doi.org/10.1089/ast.2015.1360
[132] Powner, M. W., Gerland, B., & Sutherland, J. D. (2009). Synthesis of activated pyrimidine ribonucleotides in prebiotically plausible conditions. *Nature*, 459(7244), 239-242. https://doi.org/10.1038/nature08013
[133] Jarosik, N., Bennett, C. L., Dunkley, J., Gold, B., Greason, M. R., Halpern, M., Hill, R. S., Hinshaw, G., Kogut, A., Komatsu, E., Larson, D., Limon, M., Meyer, S. S., Nolta, M. R., Odegard, N., Page, L., Smith, K. M., Spergel, D. N., Tucker, G. S., Weiland, J. L., Wollack, E., & Wright, E. L. (2011). Seven-year Wilkinson Microwave Anisotropy Probe (WMAP) observations: Sky maps, systematic errors, and basic results. *The Astrophysical Journal Supplement Series*, 192(2), Article 14. https://doi.org/10.1088/0067-0049/192/2/14
[134] Lunine, J. I. (1999). *Earth: Evolution of a habitable world*. Cambridge University Press. ISBN: 0521644232.
[135] BBC News. (2009, August 18). 'Life chemical' detected in comet. http://news.bbc.co.uk/2/hi/8208307.stm
[136] Lepot, K., Benzerara, K., Brown, G. E., & Philippot, P. (2008). Microbially influenced formation of 2.7 billion-year-old stromatolites. *Nature Geoscience*, 1(2), 118–121. https://doi.org/10.1038/ngeo107
[137] Pearce, B. K. D., Tupper, A. S., Pudritz, R. E., & Higgs, P. G. (2018). Constraining the time interval for the origin of life on Earth. *Astrobiology*, 18(3), 343–364. https://doi.org/10.1089/ast.2017.1674
[138] Forterre, P. (2005). The two ages of the RNA world, and the transition to the DNA world: A story of viruses and cells. *Biochimie*, 87(9–10), 793–803. https://doi.org/10.1016/j.biochi.2005.03.015
[139] Forterre, P. (2006). The origin of viruses and their possible roles in major evolutionary transitions. *Virus Research*, 117(1), 5–16. https://doi.org/10.1016/j.virusres.2006.01.010
[140] Warmflash, D., & Weiss, B. (2005). Did life come from another world? *Scientific American*, 293(5), 64–71. https://doi.org/10.1038/scientificamerican1105-64

[141] Witt, A. N., Vijh, U. P., & Gordon, K. D. (2003). Discovery of blue fluorescence by polycyclic aromatic hydrocarbon molecules in the Red Rectangle. *Bulletin of the American Astronomical Society*, 35, 1381. https://ui.adsabs.harvard.edu/abs/2003AAS...20311017W/abstract
[142] Battersby, S. (2004, January 11). Space molecules point to organic origins. *New Scientist*. http://www.newscientist.com/article/dn4552-space-molecules-point-to-organic-origins.html
[143] Dyson, F. J. (1999). *Origins of life* (2nd ed.). Cambridge University Press. ISBN: 9780521626682.
[144] Margulis, L. (1970). *Origin of eukaryotic cells: Evidence and research implications for a theory of the origin and evolution of microbial, plant, and animal cells on the Precambrian Earth*. Yale University Press. ISBN: 9780300013539.
[145] Margulis, L. (1981). *Symbiosis in cell evolution: Life and its environment on the early Earth*. W. H. Freeman. ISBN: 9780716712569.
[146] Margulis, L., & Sagan, D. (1995). *What is life?*. Simon & Schuster. ISBN: 9780684810164.
[147] Michalski, J. R., & Noe Dobrea, E. Z. (2007). Evidence for a sedimentary origin of clay minerals in the Mawrth Vallis region, Mars. *Geology*, 35(10), 951–954. ADS+3AGU Publications+3marsnext.jpl.nasa.gov+3
[148] NASA. (2008). NASA Orbiter Finds Martian Rock Record With 10 Beats to the Bar. *NASA Jet Propulsion Laboratory News*. NASA Jet Propulsion Laboratory (JPL)
[149] Amils, R., González-Toril, E., Aguilera, Á., Rodríguez, N., Fernández-Remolar, D., Gómez, F., ... & Sanz, J. L. (2011). From Río Tinto to Mars: The terrestrial and extraterrestrial ecology of acidophiles. *Advances in Applied Microbiology*, 77, 41–70. ResearchGate
[150] Koonin, E. V., & Dolja, V. V. (2006). Evolution of complexity in the viral world: The dawn of a new vision. *Virus Research*, 117(1), 1–4. PubMed
[151] Wei, R., Martin, T. G., Rant, U., & Dietz, H. (2012). DNA origami gatekeepers for solid-state nanopores. *Angewandte Chemie International Edition*, 51(20), 4864–4867.
[152] Dobzhansky, T. (1970). *Genetics of the evolutionary process*. Columbia University Press. AbeBooks
[153] Darwin, C. (1859). *On the origin of species by means of natural selection, or the preservation of favoured races in the struggle for life*. John Murray. darwin-online.org.uk

[154] Dawkins, R. (1996). *The blind watchmaker: Why the evidence of evolution reveals a universe without design*. W. W. Norton & Company. Amazon+1Home Page+1
[155] Chiao, J. Y., & Blizinsky, K. D. (2010). Culture–gene coevolution of individualism–collectivism and the serotonin transporter gene. *Proceedings of the Royal Society B: Biological Sciences*, 277(1681), 529–537.
[156] Goonatilake, S. (1991). *The evolution of information: Lineages in gene, culture, and artefact*. Pinter Publishers. AbeBooks
[157] Bell, P. J. L. (2001). Viral eukaryogenesis: Was the ancestor of the nucleus a complex DNA virus? *Journal of Molecular Evolution*, 53(3), 251–256.
[158] d'Herelle, F. (1922). *The bacteriophage: Its role in immunity*. Baltimore, MD: Williams & Wilkins.
[159] Haldane, J. B. S. (1929). The origin of life. *The Rationalist Annual*, 148, 3-10.
[160] McClintock, B. (1950). The origin and behavior of mutable loci in maize. *Proceedings of the National Academy of Sciences*, 36(6), 344-355. https://doi.org/10.1073/pnas.36.6.344
[161] Gilbert, W. (1986). Origin of life: The RNA world. *Nature*, 319(6055), 618. https://doi.org/10.1038/319618a0
[162] Wächtershäuser, G. (1988). Before enzymes and templates: Theory of surface metabolism. *Microbiological Reviews*, 52(4), 452-484. https://doi.org/10.1128/mr.52.4.452-484.1988
[163] Arnaud, F., Caporale, M., Varela, M., Biek, R., Chessa, B., Alberti, A., Golder, M., Mura, M., Zhang, Y., Yu, L., Pereira, G. M., & Palmarini, M. (2007). A paradigm for virus–host coevolution: Sequential counter-adaptations between endogenous and exogenous retroviruses. *PLoS Pathogens*, 3(11), e170. https://doi.org/10.1371/journal.ppat.0030170
[164] Worobey, M., Bjork, A., & Wertheim, J. O. (2007). Point, counterpoint: The evolution of pathogenic viruses and their human hosts. *Annual Review of Ecology, Evolution, and Systematics*, 38(1), 515-540. https://doi.org/10.1146/annurev.ecolsys.38.091206.095722
[165] Lobo, F. P., Mota, B. E. F., Pena, S. D. J., Azevedo, V., Macedo, A. M., Tauch, A., & Rocha, E. P. C. (2009). Virus-host coevolution: Common patterns of nucleotide motif usage in Flaviviridae and their hosts. *PLoS ONE*, 4(7), e6282. https://doi.org/10.1371/journal.pone.0006282
[166] Smith, S. E. P., Li, J., Garbett, K., Mirnics, K., & Patterson, P. H. (2007). Maternal immune activation alters fetal brain development

through interleukin-6. *Journal of Neuroscience*, 27(40), 10695-10702. https://doi.org/10.1523/JNEUROSCI.2178-07.2007
[167] Brown, A. S. (2006). Prenatal infection as a risk factor for schizophrenia. *Schizophrenia Bulletin*, 32(2), 200-202. https://doi.org/10.1093/schbul/sbj052
[168] Brown, A. S., & Susser, E. S. (2002). In utero infection and adult schizophrenia. *Mental Retardation and Developmental Disabilities Research Reviews*, 8(1), 51-57. https://doi.org/10.1002/mrdd.10004
[169] Brown, A. S., Begg, M. D., Gravenstein, S., Schaefer, C. A., Wyatt, R. J., Bresnahan, M., Babulas, V. P., & Susser, E. S. (2004). Serologic evidence of prenatal influenza in the etiology of schizophrenia. *Archives of General Psychiatry*, 61(8), 774–780. https://doi.org/10.1001/archpsyc.61.8.774
[170] Reddy, P. S., Burroughs, K. D., Hales, L. M., Ganesh, S., Jones, B. E., Idamakanti, N., Hay, C., Li, S. S., Skele, K. L., Vasko, A. J., Yang, J., Watkins, D. N., Rudin, C. M., & Hallenbeck, P. L. (2007). Seneca Valley virus, a systemically deliverable oncolytic picornavirus, and the treatment of neuroendocrine cancers. *Journal of the National Cancer Institute*, 99(21), 1623–1633. https://doi.org/10.1093/jnci/djm204
[171] Bandea, C. I. (2009). The origin and evolution of viruses as molecular organisms. *Nature Precedings*. https://doi.org/10.1038/npre.2009.3886.1
[172] Zhao, H., Gao, X., & Zhu, L. (2006). A virus co-evolution genetic algorithm based on niche technology. *2006 IEEE International Conference on Information Acquisition*, 894–899. https://doi.org/10.1109/ICIA.2006.305961
[173] Tamura, K., Mutoh, A., Nakamura, T., & Itoh, H. (2001). Linear genetic programming with virus individuals. *Proceedings of the 2nd International Workshop on Software Engineering, Artificial Intelligence, Networking and Parallel/Distributed Computing*, 936–941.
[174] Fire, A., Xu, S., Montgomery, M. K., Kostas, S. A., Driver, S. E., & Mello, C. C. (1998). Potent and specific genetic interference by double-stranded RNA in *Caenorhabditis elegans*. *Nature*, 391(6669), 806–811. https://doi.org/10.1038/35888
[175] Bernstein, E., Caudy, A. A., Hammond, S. M., & Hannon, G. J. (2001). Role for a bidentate ribonuclease in the initiation step of RNA interference. *Nature*, 409(6818), 363–366. https://doi.org/10.1038/35053110
[176] Siolas, D., Lerner, C., Burchard, J., Ge, W., Linsley, P. S., Paddison, P. J., Hannon, G. J., & Cleary, M. A. (2005). Synthetic

shRNAs as potent RNAi triggers. *Nature Biotechnology*, 23(2), 227–231. https://doi.org/10.1038/nbt1058
[177] Kanzaki, L. I. B., Ornelas, S. S., & Argañaraz, E. R. (2008). RNA interference and HIV-1 infection. *Reviews in Medical Virology*, 18(1), 5–18. https://doi.org/10.1002/rmv.553
[178] Weinberg, M. S., & Morris, K. V. (2006). Are viral-encoded microRNAs mediating latent HIV-1 infection? *DNA and Cell Biology*, 25(4), 223–231. https://doi.org/10.1089/dna.2006.25.223
[179] Yeung, M. L., Bennasser, Y., Watashi, K., Le, S. Y., Houzet, L., & Jeang, K. T. (2005). Changes in microRNA expression profiles in HIV-1-transfected human cells. *Retrovirology*, 2, Article 81. https://doi.org/10.1186/1742-4690-2-81
[180] Valeri, N., Vannini, I., Fanini, F., Calore, F., Adair, B., & Fabbri, M. (2009). Epigenetics, miRNAs, and human cancer: A new chapter in human gene regulation. *Mammalian Genome*, 20(9-10), 573–580. https://doi.org/10.1007/s00335-009-9210-0
[181] Gabuzda, D. H., Lawrence, K., Langhoff, E., Terwilliger, E., Dorfman, T., Haseltine, W. A., & Sodroski, J. (1992). Role of Vif in replication of human immunodeficiency virus type 1 in CD4+ T lymphocytes. *Journal of Virology*, 66(11), 6489–6495. https://doi.org/10.1128/JVI.66.11.6489-6495.1992
[182] Sheehy, A. M., Gaddis, N. C., Choi, J. D., & Malim, M. H. (2002). Isolation of a human gene that inhibits HIV-1 infection and is suppressed by the viral Vif protein. *Nature*, 418(6898), 646–650. https://doi.org/10.1038/nature00939
[183] Mangeat, B., Turelli, P., Liao, S., & Trono, D. (2004). A single amino acid determinant governs the species-specific sensitivity of APOBEC3G to Vif action. *Journal of Biological Chemistry*, 279(15), 14481–14483. https://doi.org/10.1074/jbc.C400060200
[184] Zhang, H., Yang, B., Pomerantz, R. J., Zhang, C., Arunachalam, S. C., & Gao, L. (2003). The cytidine deaminase CEM15 induces hypermutation in newly synthesized HIV-1 DNA. *Nature*, 424(6944), 94–98. https://doi.org/10.1038/nature01707
[185] Conticello, S. G., Thomas, C. J., Petersen-Mahrt, S. K., & Neuberger, M. S. (2005). Evolution of the AID/APOBEC family of polynucleotide (deoxy)cytidine deaminases. *Molecular Biology and Evolution*, 22(2), 367–377. https://doi.org/10.1093/molbev/msi026
[186] James, G., Witten, D., Hastie, T., & Tibshirani, R. (2013). *An Introduction to Statistical Learning: With Applications in R*. Springer. Amazon+1ACM Digital Library+1

[187] Schutz, B. F. (1980). *Geometrical Methods of Mathematical Physics*. Cambridge University Press.
[188] Gockeler, M., & Schucker, T. (1987). *Differential Geometry, Gauge Theories, and Gravity*. Cambridge University Press.
[189] Misner, C. W., Thorne, K. S., & Wheeler, J. A. (1973). *Gravitation*. W. H. Freeman. Wikipedia
[190] Bishop, R. L., & Crittenden, R. J. (1964). *Geometry of Manifolds*. Academic Press. valore.com
[191] Amari, S. (1991). Dualistic geometry of the manifold of higher-order neurons. *Neural Networks*, 4(4), 443–451. SpringerLink+3Scholars Portal Journals+3SpringerLink+3
[192] Amari, S. (1995). Information geometry of the EM and em algorithms for neural networks. *Neural Networks*, 8(9), 1379–1408.
[193] Amari, S., & Nagaoka, H. (2000). *Methods of Information Geometry* (Vol. 191). American Mathematical Society.
[194] Cencov, N. N. (1982). *Statistical Decision Rules and Optimal Inference* (Vol. 53). American Mathematical Society. AbeBooks+2biblioteca.matem.unam.mx+2Open Library+2
[195] Eguchi, S. (1983). Second order efficiency of minimum contrast estimators in a curved exponential family. *Annals of Statistics*, 11(3), 793–803.
[196] Csiszár, I., & Tusnády, G. (1984). Information geometry and alternating minimization procedures. *Statistics & Decisions*, Supplement Issue 1, 205–237. JSTOR+2web.eecs.umich.edu+2web.eecs.umich.edu+2
[197] Kivinen, J., & Warmuth, M. K. (1997). Exponentiated gradient versus gradient descent for linear predictors. *Information and Computation*, 132(1), 1–63.
[198] Warmuth, M. K., & Jagota, A. K. (1997). Continuous and discrete-time nonlinear gradient descent: Relative loss bounds and convergence. In R. Greiner & E. Boros (Eds.), *Electronic Proceedings of the 5th International Symposium on Artificial Intelligence and Mathematics*. Retrieved from http://rutcor.rutgers.edu/~amai/aimath97/Warmuth-Jagota.ps
[199] Hochreiter, S., & Schmidhuber, J. (1997). Long short-term memory. *Neural Computation*, 9(8), 1735–1780. https://doi.org/10.1162/neco.1997.9.8.1735
[200] Yao, Y., Wang, J., Ngo, C. W., Li, X., & Chua, T. S. (2019). A survey on long-range dependencies in neural networks. *arXiv preprint arXiv:1905.06693*. Retrieved from https://arxiv.org/abs/1905.06693

[201] Bengio, Y., Simard, P., & Frasconi, P. (1994). Learning long-term dependencies with gradient descent is difficult. *IEEE Transactions on Neural Networks*, 5(2), 157–166. https://doi.org/10.1109/72.279181
[202] Brown, T. B., Mann, B., Ryder, N., Subbiah, M., Kaplan, J., Dhariwal, P., ... & Amodei, D. (2020). Language models are few-shot learners. *arXiv preprint arXiv:2005.14165*. Retrieved from https://arxiv.org/abs/2005.14165
[203] Radford, A., Wu, J., Child, R., Luan, D., Amodei, D., & Sutskever, I. (2019). Language models are unsupervised multitask learners. *OpenAI Blog*, 1(8), 9. Retrieved from https://cdn.openai.com/better-language-models/language_models_are_unsupervised_multitask_learners.pdf
[204] Tay, Y., Dehghani, M., Bahri, D., & Metzler, D. (2020). Efficient transformers: A survey. *arXiv preprint arXiv:2009.06732*. Retrieved from https://arxiv.org/abs/2009.06732
[205] Kitaev, N., Kaiser, Ł., & Levskaya, A. (2020). Reformer: The efficient transformer. *arXiv preprint arXiv:2001.04451*. Retrieved from https://arxiv.org/abs/2001.04451
[206] Beltagy, I., Peters, M. E., & Cohan, A. (2020). Longformer: The long-document transformer. *arXiv preprint arXiv:2004.05150*. Retrieved from https://arxiv.org/abs/2004.05150
[207] Steriade, M., & McCarley, R. W. (2005). *Brain control of wakefulness and sleep*. Springer Science & Business Media.
[208] Posner, M. I., & Petersen, S. E. (1990). The attention system of the human brain. *Annual Review of Neuroscience*, 13(1), 25–42. https://doi.org/10.1146/annurev.ne.13.030190.000325
[209] Rayner, K. (1998). Eye movements in reading and information processing: 20 years of research. *Psychological Bulletin*, 124(3), 372–422.
[210] Findlay, J. M., & Gilchrist, I. D. (2003). *Active Vision: The Psychology of Looking and Seeing*. Oxford University Press.
[211] Gazzaniga, M. S. (2005). Forty-five years of split-brain research and still going strong. *Nature Reviews Neuroscience*, 6(8), 653–659.
[212] Arbib, M. A. (2012). *How the Brain Got Language: The Mirror System Hypothesis*. Oxford University Press.
[213] Rogers, L. J., & Andrew, R. J. (2002). *Comparative Vertebrate Lateralization*. Cambridge University Press.
[214] Rutz, C., & St Clair, J. J. H. (2012). The evolutionary origins and ecological context of tool use in New Caledonian crows. *Behavioural Processes*, 89(2), 153–165.
[215] Nottebohm, F. (1972). The origins of vocal learning. *The American Naturalist*, 106(947), 116–140.

[216] Alvarez-Buylla, A., & Nottebohm, F. (1988). Migration of young neurons in adult avian brain. *Nature*, 335(6188), 353–354.
[217] Fagot, J., & Vauclair, J. (1991). Manual laterality in nonhuman primates: A distinction between handedness and manual specialization. *Psychological Bulletin*, 109(1), 76–89.
[218] Rogers, L. J., & Andrew, R. J. (2002). *Comparative Vertebrate Lateralization*. Cambridge University Press.
[219] Bisazza, A., Cantalupo, C., Capocchiano, M., & Vallortigara, G. (2000). Population lateralization and social behaviour: A study with 16 species of fish. *Laterality: Asymmetries of Body, Brain and Cognition*, 5(3), 269–284.
[220] Fragaszy, D. M., & Visalberghi, E. (2004). Socially biased learning in monkeys. *Learning & Behavior*, 32(1), 24–35.
[221] Hopkins, W. D., Pilcher, D., & Cantalupo, C. (2004). Neural basis of cognition, communication and handedness in nonhuman primates. In D. V. M. Bishop & S. A. D. Watkins (Eds.), *Neurobiology of Language* (pp. 35-46). Cambridge University Press.
[222] Rilling, J. K., & Insel, T. R. (1999). The primate neocortex in comparative perspective using magnetic resonance imaging. *Journal of Human Evolution*, 37(2), 191-223. https://doi.org/10.1006/jhev.1999.0313
[223] Arbib, M. A. (2009). Evolving the language-ready brain and the social mechanisms that support language. *Journal of Communication Disorders*, 42(4), 263-271. https://doi.org/10.1016/j.jcomdis.2009.03.009
[224] Corballis, M. C. (2009). The evolution of language. *Annals of the New York Academy of Sciences*, 1156(1), 19-43. https://doi.org/10.1111/j.1749-6632.2009.04423.x
[225] Cantalupo, C., & Hopkins, W. D. (2001). Asymmetric Broca's area in great apes. *Nature*, 414(6863), 505. https://doi.org/10.1038/35107134
[226] Johanson, D. C., & Edgar, B. (1996). *From Lucy to language*. Simon & Schuster.
[227] Wood, B., & Richmond, B. G. (2000). Human evolution: Taxonomy and paleobiology. *Journal of Anatomy*, 197(1), 19-60. https://doi.org/10.1046/j.1469-7580.2000.19710019.x
[228] Smithsonian Institution. (n.d.). *Homo habilis*. The Smithsonian Institution's Human Origins Program. Retrieved April 9, 2025, from https://humanorigins.si.edu/evidence/human-fossils/species/homo-habilis
[229] Tattersall, I. (1998). *Becoming human: Evolution and human uniqueness*. Houghton Mifflin Harcourt.
[230] Wrangham, R. (2009). *Catching fire: How cooking made us human*. Basic Books.

[231] Rightmire, G. P. (2004). Brain size and encephalization in early to Mid-Pleistocene Homo. *American Journal of Physical Anthropology*, 124(2), 109-123. https://doi.org/10.1002/ajpa.10346
[232] Tattersall, I., & Schwartz, J. H. (2000). *Extinct humans*. Westview Press.
[233] Soressi, M., & Hublin, J.-J. (2013). The Mousterian of the French Middle Paleolithic: A Review. *Quaternary International*, 295, 1–36. https://doi.org/10.1016/j.quaint.2012.07.018
[234] Bar-Yosef, O., & Belfer-Cohen, A. (2001). From Africa to Eurasia—Early Dispersals. *Quaternary International*, 75(1), 19–28. https://doi.org/10.1016/S1040-6182(00)00075-8
[235] Trinkaus, E. (1995). Neanderthal Mortality Patterns. *Journal of Archaeological Science*, 22(1), 121–142. https://doi.org/10.1016/S0305-4403(95)80170-7
[236] Conde-Valverde, M., Martínez, I., Quam, R., Rosa, M., Lorenzo, C., & Jarabo, P. (2024). The Child Who Lived: Down Syndrome Among Neanderthals? *Science Advances*, 10(26), eadn9310. https://doi.org/10.1126/sciadv.adn9310
[237] Trinkaus, E. (1995). Neanderthal Mortality Patterns. *Journal of Archaeological Science*, 22(1), 121–142. https://doi.org/10.1016/S0305-4403(95)80170-7
[238] Pettitt, P. (2002). The Neanderthal Dead: Exploring Mortuary Variability in Middle Palaeolithic Eurasia. *Before Farming*, 2002(1), 1–26. https://doi.org/10.3828/bfarm.2002.1.1
[239] Harari, Y. N. (2015). *Sapiens: A Brief History of Humankind*. Harper. https://www.harpercollins.com/products/sapiens-yuval-noah-harari
[240] Mellars, P. (1996). *The Neanderthal Legacy: An Archaeological Perspective from Western Europe*. Princeton University Press. https://press.princeton.edu/books/paperback/9780691167985/the-neanderthal-legacy
[241] Green, R. E., Krause, J., Briggs, A. W., Maricic, T., Stenzel, U., Kircher, M., Patterson, N., Li, H., Zhai, W., Fritz, M. H.-Y., Hansen, N. F., Durand, E. Y., Malaspinas, A.-S., Jensen, J. D., Marques-Bonet, T., Alkan, C., Prüfer, K., Meyer, M., Burbano, H. A., … Pääbo, S. (2010). A Draft Sequence of the Neandertal Genome. *Science*, 328(5979), 710–722. https://doi.org/10.1126/science.1188021
[242] Pääbo, S. (2014). *Neanderthal Man: In Search of Lost Genomes*. Basic Books. https://www.basicbooks.com/titles/svante-paabo/neanderthal-man/9780465080687/

[243] Reich, D., Green, R. E., Kircher, M., Krause, J., Patterson, N., Durand, E. Y., Viola, B., Briggs, A. W., Stenzel, U., Johnson, P. L. F., Maricic, T., Good, J. M., Marques-Bonet, T., Alkan, C., Fu, Q., Mallick, S., Li, H., Meyer, M., Eichler, E. E., Stoneking, M., … Pääbo, S. (2010). Genetic History of an Archaic Hominin Group from Denisova Cave in Siberia. *Nature*, 468(7327), 1053–1060. https://doi.org/10.1038/nature09710
[244] Meyer, M., Kircher, M., Gansauge, M.-T., Li, H., Racimo, F., Mallick, S., Schraiber, J. G., Jay, F., Prüfer, K., de Filippo, C., Sudmant, P. H., Alkan, C., Fu, Q., Do, R., Rohland, N., Schreiber, S., Sawyer, S., Green, R. E., Bryc, K., … Pääbo, S. (2012). A High-Coverage Genome Sequence from an Archaic Denisovan Individual. *Science*, 338(6104), 222–226. https://doi.org/10.1126/science.1224344
[245] Derevianko, A. P., Shunkov, M. V., & Volkov, P. V. (2008). A Paleolithic bracelet from Denisova Cave. *Archaeology, Ethnology & Anthropology of Eurasia*, 34(2), 13–25. https://doi.org/10.1016/j.aeae.2008.08.001
[246] Zilhão, J., & d'Errico, F. (1999). The chronology and taphonomy of the earliest Aurignacian and its implications for the understanding of Neandertal extinction. *Journal of World Prehistory*, 13, 1–68. https://doi.org/10.1023/A:1022348410845
[247] Karmin, M., Saag, L., Vicente, M., Sayres, M. A. W., Järve, M., Talas, U. G., ... & Villems, R. (2015). A recent bottleneck of Y chromosome diversity coincides with a global change in culture. *Genome Research*, 25(4), 459–466. https://doi.org/10.1101/gr.186684.114
[248] Hublin, J.-J. (2009). The origin of Neandertals. *Proceedings of the National Academy of Sciences*, 106(38), 16022–16027. https://doi.org/10.1073/pnas.0904119106
[249] Lieberman, P. (2007). The evolution of human speech: Its anatomical and neural bases. *Current Anthropology*, 48(1), 39–66. https://doi.org/10.1086/509092
[250] Fitch, W. T. (2010). *The evolution of language*. Cambridge University Press. https://doi.org/10.1017/CBO9780511817779
[251] Tomasello, M. (1999). *The cultural origins of human cognition*. Harvard University Press. https://doi.org/10.4159/9780674026469
[252] McBrearty, S., & Brooks, A. S. (2000). The revolution that wasn't: A new interpretation of the origin of modern human behavior. *Journal of Human Evolution*, 39(5), 453–563. https://doi.org/10.1006/jhev.2000.0435

[253] Henshilwood, C. S., & Marean, C. W. (2003). The origin of modern human behavior: Critique of the models and their test implications. *Current Anthropology*, 44(5), 627–651. https://doi.org/10.1086/377665
[254] Bickerton, D. (1990). *Language and species*. University of Chicago Press. https://doi.org/10.7208/chicago/9780226046136.001.0001
[255] Corballis, M. C. (2002). *From hand to mouth: The origins of language*. Princeton University Press. https://doi.org/10.1515/9781400841653
[256] Mithen, S. (2005). *The singing Neanderthals: The origins of music, language, mind, and body*. Weidenfeld & Nicolson. https://doi.org/10.5860/choice.43-2729
[257] Dunbar, R. I. M. (1998). *Grooming, Gossip, and the Evolution of Language*. Harvard University Press.
[258] Carstairs-McCarthy, A. (1999). *The Origins of Complex Language: An Inquiry into the Evolutionary Beginnings of Sentences, Syllables, and Truth*. Oxford University Press.
[259] Everett, D. L. (2017). *How Language Began: The Story of Humanity's Greatest Invention*. Liveright Publishing.
[260] Honing, H. (Ed.). (2018). *The Origins of Musicality*. MIT Press.
[261] Turk, M., Turk, I., & Otte, M. (2020). The Neanderthal Musical Instrument from Divje Babe I Cave (Slovenia): A Critical Review of the Discussion. *Applied Sciences*, 10(4), 1226. https://doi.org/10.3390/app10041226
[262] Cross, I. (2001). Music, Mind and Evolution. *Psychology of Music*, 29(1), 95–102. https://doi.org/10.1177/0305735601291007
[263] Nettl, B. (2005). *The Study of Ethnomusicology: Thirty-One Issues and Concepts* (2nd ed.). University of Illinois Press.
[264] Pinker, S. (1997). *How the Mind Works*. W. W. Norton & Company.
[265] Doty, D. B. (2004). My Role in the Revival of Just Intonation. *1/1, The Journal of the Just Intonation Network*, 17(1). Retrieved from https://www.dbdoty.com/Words/Doty_one-one_Spring_04.pdf
[266] Garcia-Ballester, Luis. Galen and Galenism, Burlington: Ashgate-Variorum, p. 1663. (2002).
[267] Eichholz, D.E. Galen and His Environment, Greece & Rome 20 no. 59, Cambridge University Press, pp. 60–71. (1951).
[268] Psychoyos, Dimitris K. The forgotten art of isopsephy and the magic number KZ. Semiotica 154 (1–4): 157–224. (April 2005).
[269] Healey, J. F. (1990). *The Early Alphabet*. In *The Cambridge Encyclopedia of the World's Ancient Languages* (pp. 1-20). Cambridge University Press.

[270] Assyrian homa, or sacred tree; Tree of life, Assyrian. (1897). *The New York Public Library Digital Collections*. Retrieved April 9, 2025, from https://digitalcollections.nypl.org/items/510d47e1-165e-a3d9-e040-e00a18064a99NYPL Digital Collections
[271] Horsley, Richard A. (2003). "The Message of the Tree of Life: A Study of the Tree of Life in the Hebrew Bible." *Journal of Biblical Literature*, 122(2), 233-250.
[272] Horsley, Richard A. (2003). "Silphium: The Plant That Was Too Good to Last." *The Classical World*, 97(3), 273-284.
[273] Balogh, C. (2024). The tree of life in ancient Near Eastern iconography. *OnScript*. Retrieved April 9, 2025, from https://onscript.study/wp-content/uploads/2024/11/Balogh-Tree-of-Life.pdfOnScript
[274] Kaplan, A. (1990). *Sefer Yetzirah: The book of creation in theory and practice*. Samuel Weiser.Alljudaica+1Yesterday's Muse Books+1
[275] Dan, J. (2006). *Kabbalah: A very short introduction*. Oxford University Press.Goodreads
[276] No reference found for "The Tree of Life: An Exploration of Biblical Symbolism" by David L. Smith.
[277] Katz, S. T. (Ed.). (1983). *Mysticism and religious traditions*. Oxford University Press.Amazon+3IMERE+3Oxford University Press+3
[278] Matt, D. C. (1995). *The essential Kabbalah: The heart of Jewish mysticism*. HarperSanFrancisco.
[279] No reference found for "Kabbalistic Mysticism: A Study of the Tree of Life" by Moshe Idel.
[280] Steinberg, D. A. (n.d.). *Gematria: A New Interpretation*. [No additional information available.]
[281] Cooper, D. A. (2005). *God Is a Verb: Kabbalah and the Practice of Mystical Judaism*. Riverhead Books. Google Books+1SoBrief+1
[282] Gordon, R. (1990). *Image and Value in the Greco-Roman World: Studies in Mithraism and Religious Art*. Variorum.Roman religion
[283] McGinnis, J. F. (2014). *Divination and the Politics of the Future in Ancient Rome*. [No additional information available.]
[284] Greer, J. M. (2009). *The Art and Practice of Geomancy: Divination, Magic, and Earth Wisdom of the Renaissance*. Weiser Books. Red Wheel Weiser+8Everand+8Green Hand Bookshop+8
[285] Place, R. M. (2009). *The Tarot: History, Symbolism, and Divination*. TarcherPerigee.
[286] Dean, L. (2019). *The Divination Handbook: The Modern Seer's Guide to Using Tarot, Crystals, Palmistry, and More*. Fair Winds Press. Liz Dean+4Barnes & Noble+4Goodreads+4

[287] Greer, J. M. (2012). *The Art and Practice of Geomancy: Divination, Magic, and Earth Wisdom of the Renaissance*. Weiser Books.
[288] Eason, C. (2007). *Scrying the Secrets of the Future: How to Use Crystal Ball, Fire, Wax, Mirrors, Shadows, and Spirit Guides to Reveal Your Destiny*. New Page Books. Barnes & Noble+6Internet Archive+6Nowhere Bookshop+6
[289] Nelson, J. (2008). *The Magic Mirror: Divination Through the Ancient Art of Scrying*. Hampton Roads Publishing. Amazon+1Amazon+1
[290] Keightley, D. N. (1978). *Sources of Shang History: The Oracle-Bone Inscriptions of Bronze Age China*. University of California Press. JSTOR+4Wikipedia+4Internet Archive+4
[291] Goldin, P. R. (2005). *After Confucius: Studies in Early Chinese Philosophy*. University of Hawaii Press.
[292] Freud, S. (1899). *The Interpretation of Dreams*. Vienna: Franz Deuticke.
[293] Wilhelm, R. (1967). *I Ching: The Book of Changes* (C. F. Baynes, Trans.). Princeton University Press.
[294] Legge, J. (1963). *The I Ching: Book of Changes*. Dover Publications.AbeBooks+15Architectural Digest+15Freud Museum London+15
[295] Barnes, M. P. (2012). *Runes: A Handbook*. Boydell Press.
[296] Blum, R. (1982). *The Book of Runes: A Handbook for the Use of an Ancient Oracle: The Viking Runes*. St. Martin's Press.Regesta Imperii OPAC+3Chamblin Bookmine+3Internet Archive+3
[297] Becker, Andrea. "The Royal Game of Ur,". In Ancient Board Games in Perspective: Papers from the 1990 British Museum Colloquium, with Additional Contributions. London, England: British Museum Press, 2007.
[298] Sullivan, J. (2011). "Playing with the Past: The Archaeology of Board Games." *World Archaeology*, 43(3), 345-360.
[299] Pingree, D. (1997). *Astral Magic in Medieval Islam and the Christian West: Platonizing the Astrological Imagination*. Brill.JSTOR
[300] Kafadar, C. (1996). *Between Two Worlds: The Construction of the Ottoman State*. University of California Press.
[301] Jeffrey Bennett; Megan Donohue; Nicholas Schneider; Mark Voit (2007). The cosmic perspective (4th ed.). San Francisco, CA: Pearson/Addison-Wesley. pp. 82–84.

[302] Rutkin, H. Darell (2006). "Astrology". In K. Park; L. Daston (eds.). Early Modern Science. The Cambridge History of Science. Vol. 3. Cambridge University Press. pp. 541–561.
[303] Decker, R., & Dummett, M. (2013). *A History of the Occult Tarot*. Duckworth.
[304] Case, P. F. (2006). *The Tarot: A Key to the Wisdom of the Ages*. TarcherPerigee.
[305] Waite, A. E. (2005). *The Pictorial Key to the Tarot*. Kessinger Publishing.
[306] Matthews, Caitlin. Untold Tarot: The Lost Art of Reading Ancient Tarots. REDFeather; 1st edition, 2018.
[307] Kaplan, S. R. (1972). *Tarot Classic*. U.S. Games Systems.
[308] Fortune, D. (2000). *The Mystical Qabalah*. Weiser Books.
[309] Hanegraaff, Wouter J. New Age Religion and Western Culture: Esotericism in the Mirror of Secular Thought. SUNY Press. p. 396. (1998).
[310] Stratton-Kent, Jake. "The English Qaballa". The Equinox: British Journal of Thelema. VII (1): 17–25. (March 1988).
[311] Stratton-Kent, Jake. " What is a Qabalah?". The Equinox: British Journal of Thelema. VII (2): 59-61. (May 1988b).
[312] Ibn Arabi. (1980). *The Bezels of Wisdom* (R. W. J. Austin, Trans.). Paulist Press.
[313] Abad, Jose. Tonos Digital; Numero 15 – Junio 2008. "La virtu segun maquiavelo: Significados Y Traducciones."
[314] Skinner, Q. (2000). *Machiavelli: A Very Short Introduction*. Oxford University Press.
[315] Coleridge's Philosophy: The Logos as Unifying Principle. Clarendon Press. ISBN: 978-0198240754.
[316] S.T Coleridge, "On the Principles of Genial Criticism", Felix Farley's Bristol Journal, 27 August 1814.
[317] Thomas McFarland Ed. The Collected Works of Samuel Taylor Coleridge, Volume 15: Opus Maximum. 2002. ISBN: 9780691098821
[318] Kiparsky, Paul. "Stress, Syntax, and Meter", Language, 51 (3): 576-616.
[319] Cherry, J. F. (Ed.). (2016). *Prehistory in the Balkans: Interregional Research and Educational Practices*. Center for Prehistoric Research.
[320] Koromela, M. (2002). *The Greeks in the Black Sea: From the Bronze Age to the Early Twentieth Century* (2nd ed.). Maniateas.
[321] King, C. (2004). *The Black Sea: A History*. Oxford University Press.

[322] Cowan, C. W., & Watson, P. J. (Eds.). (2006). *The Origins of Agriculture: An International Perspective*. University of Alabama Press.
[323] Whitehouse, D. (1997). *Glass of the Roman Empire*. Corning Museum of Glass.
[324] Fischer, P. M. (2014). *The Archaeology of the Bronze Age in Cyprus: Studies in Mediterranean Archaeology*. Åström Editions.
[325] Anthony, D. W. (2007). *The Horse, the Wheel, and Language: How Bronze-Age Riders from the Eurasian Steppes Shaped the Modern World*. Princeton University Press.
[326] Cherry, J. F. (Ed.). (2016). *Prehistory in the Balkans: Interregional Research and Educational Practices*. Center for Prehistoric Research.
[327] Schmidt, K. (2011). Göbekli Tepe: The World's First Temple? *Scientific American*, 304(4), 40-47.
[328] Balter, M. (2010). The birth of religion. *Science*, 328(5981), 1148-1150. https://doi.org/10.1126/science.328.5981.1148
[329] Hodder, I. (2020). *Göbekli Tepe: The World's First Temple*. Cambridge University Press.
[330] Mellaart, J. (1967). *Çatal Hüyük: A Neolithic Town in Anatolia*. McGraw-Hill.
[331] Kenyon, K. M. (1960). *Excavations at Jericho*. British School of Archaeology in Jerusalem.
[332] Bar-Yosef, O., & Valla, F. R. (Eds.). (1991). *The Natufian Culture in the Levant*. International Monographs in Prehistory.
[333] Zangger, E. (1992). *The Flood from Heaven: Deciphering the Atlantis Legend*. William Morrow & Co.
[334] Finkel, I. L. (2014). *The Ark Before Noah: Decoding the Story of the Flood*. Hodder & Stoughton.
[335] Childe, V. G. (1951). *Man Makes Himself*. New American Library.
[336] Christian, D. (Ed.). (2015). *The Cambridge World History: Volume 1, Introducing World History, to 10,000 BCE*. Cambridge University Press.
[337] Daniels, P. T., & Bright, W. (Eds.). (1996). *The World's Writing Systems*. Oxford University Press.
[338] Postgate, J. N. (1992). *Early Mesopotamia: Society and Economy at the Dawn of History*. Routledge.
[339] Finley, M. I. (1999). *The Ancient Economy* (Updated ed.). University of California Press.
[340] Gies, J., & Gies, F. (1994). *Life in a Medieval Castle*. HarperCollins.

[341] Burckhardt, J. (1990). *The Civilization of the Renaissance in Italy* (S. G. C. Middlemore, Trans.). Harper & Row. (Original work published 1860)
[342] Hobsbawm, E. J. (1996). *The Age of Revolution: Europe 1789–1848*. Vintage.
[343] Eisenstein, E. L. (1983). *The Printing Revolution in Early Modern Europe*. Cambridge University Press.
[344] Ashton, T. S. (1997). *The Industrial Revolution, 1760–1830.* Oxford University Press. (Original work published 1948)
[345] Thompson, E. P. (1963). *The Making of the English Working Class*. Gollancz.
[346] Hobsbawm, E. J. (1994). *The Age of Extremes: The Short Twentieth Century, 1914–1991*. Michael Joseph.
[347] Castells, M. (1996). *The Rise of the Network Society*. Blackwell Publishers.
[348] Standage, T. (1998). *The Victorian Internet: The Remarkable Story of the Telegraph and the Nineteenth Century's On-Line Pioneers*. Walker & Company.
[349] McLuhan, M. (1964). *Understanding Media: The Extensions of Man*. McGraw-Hill.
[350] Harari, Y. N. (2015). *Sapiens: A Brief History of Humankind*. Harper.
[351] Diamond, J. (1997). *Guns, Germs, and Steel: The Fates of Human Societies*. W. W. Norton.
[352] Hammurabi. (1958). The Code of Hammurabi. In J. B. Pritchard (Ed.), *The Ancient Near East: An Anthology of Texts and Pictures*. Princeton University Press.
[353] Crook, J. A. (1955). *The Twelve Tables of Roman Law*. Cambridge University Press.
[354] Jacobsen, T. (1976). *The Treasures of Darkness: A History of Mesopotamian Religion*. Yale University Press.
[355] Assmann, J. (2001). *The Search for God in Ancient Egypt*. Cornell University Press.
[356] Boyce, M. (2001). *Zoroastrians: Their Religious Beliefs and Practices* (2nd ed.). Routledge.
[357] Wittgenstein, L. (2009). *Philosophical Investigations* (P. M. S. Hacker & J. Schulte, Eds.; G. E. M. Anscombe, P. M. S. Hacker, & J. Schulte, Trans.; Rev. 4th ed.). Wiley-Blackwell.
[358] de Saussure, F. (1966). *Course in General Linguistics* (C. Bally & A. Sechehaye, Eds.; W. Baskin, Trans.). McGraw-Hill.

[359] Lakoff, G., & Johnson, M. (2003). *Metaphors We Live By* (2nd ed.). University of Chicago Press.
[360] Bakhtin, M. M. (1981). *The Dialogic Imagination: Four Essays* (M. Holquist, Ed.; C. Emerson & M. Holquist, Trans.). University of Texas Press.
[361] Derrida, J. (1998). *Of Grammatology* (Corrected ed.; G. C. Spivak, Trans.). Johns Hopkins University Press.
[362] Frankl, V. E. (2006). *Man's Search for Meaning*. Beacon Press.
[363] Dennett, D. C. (1991). *Consciousness Explained*. Little, Brown and Company.
[364] Rorty, Richard. Philosophy and the Mirror of Nature. Princeton University Press, 1979. [365] Sapir, Edward. Language: An Introduction to the Study of Speech. Harcourt, Brace and Company, 1921.
[366] Whorf, Benjamin Lee. Language, Thought, and Reality: Selected Writings of Benjamin Lee Whorf. Edited by John B. Carroll, MIT Press, 1956.
[367] Orwell, George. 1984. Harcourt, Brace & Company, 1949.
[368] Leibniz, G. W. New Essays on Human Understanding. Edited by Peter Remnant and Jonathan Bennett, Cambridge University Press, 1981. (Originally published in 1704.)
[369] Nietzsche, Friedrich. On Truth and Lies in a Nonmoral Sense. In Philosophical Writings, edited by R. J. Hollingdale, Penguin Books, 1990.
[370] Wittgenstein, Ludwig. Tractatus Logico-Philosophicus. Translated by D. F. Pears and B. F. McGuinness, Routledge, 2001. (Originally published in 1921.)
[371] Chomsky, Noam. Language and Mind. Harcourt Brace Jovanovich, 1972.
[372] Bohm, David. On Dialogue. Routledge, 1996.
[373] Gardner, Howard. Frames of Mind: The Theory of Multiple Intelligences. Basic Books, 1983.
[374] Pollack, R. (1980). 78 Degrees of Wisdom: A Tarot Journey to Self-Awareness. This book explores the meanings of tarot cards and their applications in personal growth and storytelling.
[375] Greer, M. K. (2002). *The Tarot: History, Symbolism, and Divination*. HarperSanFrancisco.
[376] Cossu, A. (2019). *Tarot and the Archetypal Journey: The Jungian Path from Darkness to Light*. Weiser Books.
[377] Bunning, J. (1998). *Learning the Tarot: A Tarot Book for Beginners*. Weiser Books.

[378] Banzhaf, H. (2000). *Tarot and the Journey of the Hero*. Weiser Books.
[379] Kenner, C. (2009). *Tarot for Writers*. Llewellyn Publications.
[380] Kenner, C. (2009). *Tarot for Writers*. Llewellyn Publications.
[381] Stueart, J. (2024). "7 Ways Tarot Can Inspire and Revive Your Writing." *Jerome Stueart's Blog*.
[382] Katz, S. (2019). "Tarot as a Narrative Tool in Role-Playing Games." *Tabletop Gaming Magazine*, 12(3), 45-50.
[383] Diacon-Furtado, Q. (2024). "Using Tarot in the Writing Process: Leveraging Tarot Readings as a Creative Catalyst." *Writer's Digest*.
[384] Pippin Mizzi, C. (2024). *Tarot for Creativity: A Guide for Igniting Your Creative Practice*. Chronicle Books.
[385] Sullivan, M. (2021). "Tarot as a Creative Prompt: Unlocking Your Imagination." *Creativity Journal*, 15(2), 56-62.
[386] Cohen, R. (2022). The Archetypes of Tarot: A Narrative Framework. *Journal of Creative Writing Studies*, 7(1), 1-15.
[387] Davis, L. (2022). Tarot and the Art of Storytelling: Techniques for Writers. *The Creative Writer's Handbook*, 9(3), 22-30.
[388] Harris, J. (2022). Tarot as a Narrative Generator: Techniques for Writers. *The Writer's Guide*, 14(2), 45-50.
[389] Miller, T. (2022). *The Tarot Storyteller: Using Tarot Cards to Create Compelling Narratives*.
[390] Fitzgerald, A. (2023). The Role of Archetypes in Tarot and Storytelling. *Journal of Mythic Studies*, 8(1), 10-20.
[391] Lindsey, K. (2023). Tarot and the Creative Process: A Guide for Artists and Writers. *Artistic Expression Journal*, 11(4), 30-35.
[392] Rosen, M. (2023). Using Tarot for Plot Development: A Practical Approach. *The Fiction Writer's Workshop*, 6(2), 15-25.
[393] Thompson, R. (2023). Tarot as a Tool for Narrative Exploration. *Creative Writing Studies Journal*, 9(3), 50-60.
[394] VanderKam, J. C. (2012). *1 Enoch: The Hermeneia Translation*. Fortress Press. booksofm.substack.com+12Fortress Press+12Amazon+12
[395] Jung, C. G. (1981). *The Archetypes and the Collective Unconscious* (R. F. C. Hull, Trans.; 2nd ed.). Princeton University Press. City Lights+3Goodreads+3Taylor & Francis+3
[396] Jung, C. G. (1971). *Psychological Types* (R. F. C. Hull, Trans.; 2nd ed.). Princeton University Press. Amazon
[397] Hillman, J. (1996). *The Soul's Code: In Search of Character and Calling*. Random House.
[398] Tarnas, R. (2006). *Cosmos and Psyche: Intimations of a New World View*. Viking.

[399] Kalsched, D. (1996). *The Inner World of Trauma: Archetypal Defenses of the Personal Spirit*. Routledge.
[400] Stein, M. (1998). *Jung's Map of the Soul: An Introduction*. Open Court.
[401] Samuels, A. (1993). *The Political Psyche*. Routledge.
[402] James, W. (1890). *The Principles of Psychology*. Holt.
[403] Woolf, V. (1925). *Mrs. Dalloway*. Hogarth Press.
[404] Ekman, P. (1992). An argument for basic emotions. *Cognition and Emotion*, 6(3), 169–200.
[405] Bruner, J. (1990). *Acts of Meaning*. Harvard University Press.
[406] Kahneman, D. (2011). *Thinking, Fast and Slow*. Farrar, Straus and Giroux.
[407] Posner, M. I. (1980). Orienting of attention. *Quarterly Journal of Experimental Psychology*, 32(1), 3–25.
[408] Penrose, R. (1994). *Shadows of the Mind: A Search for the Missing Science of Consciousness*. Oxford University Press.
[409] Hameroff, S., & Penrose, R. (1996). Orchestrated reduction of quantum coherence in brain microtubules: A model for consciousness. *Mathematics and Computers in Simulation*, 40(3-4), 453-480. https://doi.org/10.1016/0378-4754(96)80476-9
[410] Bohm, D. (1980). *Wholeness and the Implicate Order*. Routledge. https://www.routledge.com/Wholeness-and-the-Implicate-Order/Bohm/p/book/9780415289795
[411] Tegmark, M. (2000). Importance of quantum decoherence in brain processes. *Physical Review E*, 61(4), 4194-4206. https://doi.org/10.1103/PhysRevE.61.4194
[412] Tononi, G. (2004). An information integration theory of consciousness. *BMC Neuroscience*, 5, 42. https://doi.org/10.1186/1471-2202-5-42
[413] Koch, C., Massimini, M., Boly, M., & Tononi, G. (2016). Neural correlates of consciousness: Progress and problems. *Nature Reviews Neuroscience*, 17(5), 307-321. https://doi.org/10.1038/nrn.2016.22
[414] Baars, B. J. (1988). *A Cognitive Theory of Consciousness*. Cambridge University Press.
[415] Nagel, T. (1974). What is it like to be a bat? *The Philosophical Review*, 83(4), 435-450. https://doi.org/10.2307/2183914
[416] Jackson, F. (1982). Epiphenomenal qualia. *The Philosophical Quarterly*, 32(127), 127-136. https://doi.org/10.2307/2960077
[417] Chalmers, D. J. (1996). *The Conscious Mind: In Search of a Fundamental Theory*. Oxford University Press.

[418] Baars, B. J. (1997). *In the Theater of Consciousness: The Workspace of the Mind*. Oxford University Press.
[419] Kahn, D. (2015). The psychology of consciousness. *Annual Review of Psychology*, 66, 1-25. https://doi.org/10.1146/annurev-psych-010814-015031

[420] Sartre, J.-P. (1964). *Nausea* (R. Baldick, Trans.). New Directions. (Original work published 1938).
[421] Sartre, J.-P. (1938). *La Nausée*. Gallimard.
[422] Eiseley, L. (1957). *The Immense Journey: An Imaginative Naturalist Explores the Mysteries of Man and Nature*. Random House.
[423] McGowan, J. (1957). The Immense Journey: A Review. *The New York Times Book Review*.
[424] McCarthy, J. (1976). Loren Eiseley and the Nature of Science. *The American Scholar*, 45(3), 367-375.
[425] Ockham, W. (1990). *Philosophical Writings: A Selection* (P. Boehner, Ed. & Trans.). Hackett Publishing Company. (Original work published in the 14th century).
[426] Scholem, G. (1974). *Kabbalah*. Keter Publishing House.
[427] Kaplan, A. (1982). *Meditation and Kabbalah*. Samuel Weiser, Inc.
[428] Genesis 1:26-27. The Holy Bible.
[429] Tillich, P. (1952). *The Courage to Be*. Yale University Press.
[430] Lewis, C. S. (1952). *Mere Christianity*. HarperCollins.
[431] Chomsky, N. (1965). *Aspects of the Theory of Syntax*. The MIT Press.
[432] Lewis, C. S. (2001). *The abolition of man*. HarperCollins. (Original work published 1943)
[433] Lewis, C. S. (1964). *The discarded image: An introduction to Medieval and Renaissance literature*. Cambridge University Press.
[434] Mitchell, S. (Trans.). (2000). *The Bhagavad Gita*. Three Rivers Press. *(Note: Many translations exist; this is a widely cited and accessible version.)*
[435] Lao Tzu. (2006). *Tao Te Ching* (S. Mitchell, Trans.). HarperPerennial. *(Note: Multiple respected translations exist; Stephen Mitchell's is among the most cited.)*
[436] Lennox, J. C. (2011). *God's undertaker: Has science buried God?* Lion Hudson.
[437] *Meyer, SC (2021). Return of the God Hypothesis: Three Scientific Discoveries That Reveal the Mind Behind the Universe. HarperOne. ISBN 978-0062071507.*

[438] Scholem, G. (1991). *On the Kabbalah and its symbolism* (R. Manheim, Trans.). Schocken Books. (Original work published 1960)
[439] Böhme, J. (1998). *The signature of all things: On the origin and function of the universe* (J. Ellistone, Trans.). Shambhala Publications. *(Originally written 1621)*
[440] Penrose, R., W. Rindler (1984) Volume 1: Two-Spinor Calculus and Relativistic Fields, Cambridge University Press, United Kingdom.
[441] Cailler, C. 1917. Archs. Sci. Phys. Nat. ser. 4, 44 p. 237.
[442] Girard, P.R.. The Quaternion group and modern physics. Eur. J. Phys. 5 (1984): 25-32.
[443] Synge, J.L. Quaternions, Lorentz Transformations and the Conway-Dirac-Eddington Matrices.
[444] Gaździcki, Marek; Gorenstein, Mark I. (2016), Rafelski, Johann (ed.), "Hagedorn's Hadron Mass Spectrum and the Onset of Deconfinement", Melting Hadrons, Boiling Quarks – From Hagedorn Temperature to Ultra-Relativistic Heavy-Ion Collisions at CERN, Springer International Publishing, pp. 87–92.
[445] Kato T. Fundamental Properties of Hamiltonian Operators of Schrodinger Type, Transactions of the American Mathematical Society, 1951, Pg. 195-211.
[446] Kato, T. Perturbation theory for linear operators. Springer 1980.
[447] Briggs, K. A precise calculation of the Feigenbaum constants. Mathematics of Computation, Vol. 57, Num.195, July 1991, pages 435-439.
[448] Winters-Hilt S, I. H. Redmount, and L. Parker, "Physical distinction among alternative vacuum states in flat spacetime geometries," Phys. Rev. D 60, 124017 (1999).
[449] Friedman, J.L., J. Louko, and S. Winters-Hilt. Reduced phase space formalism for spherically symmetric geometry with a massive dust shell. Phys Rev. D Vol. 56, Num 12 (1997).
[450] Louko, J. and S.N. Winters-Hilt, Phys. Rev. D 54, 2647 (1996).
[451] Louko, J., J.Z. Simon, and S.N. Winters-Hilt, Phys. Rev. D (1997)
[452] Nielsen, Frank (2022). "The Many Faces of Information Geometry". Notices of the AMS. 69 (1). American Mathematical Society: 36-45.
[453] Nielsen, Frank (2018). "An Elementary Introduction to Information Geometry". Entropy. 22 (10).
[454] Penrose, Roger (May 1996). "On Gravity's role in Quantum State Reduction". *General Relativity and Gravitation*. 28 (5): 581–600.
[455] Diósi, L. (1989). "Models for universal reduction of macroscopic quantum fluctuations". Physical Review A. 40 (3): 1165–1174.

[456] Maldacena, J.. The Large N limit of superconformal field theories and supergravity.
Advances in Theoretical and Mathematical Physics. 2 (4): 231–252.
[457] Witten, Edward (1998). "Anti-de Sitter space and holography". Advances in Theoretical and Mathematical Physics. 2 (2): 253–291.
[458] Peskin, M.E., Schroeder, D.V. (1995). An Introduction to Quantum Field Theory.
[459] Haxton, W.C.; Hamish Robertson, R.G.; Serenelli, Aldo M. (18 August 2013). "Solar Neutrinos: Status and Prospects". *Annual Review of Astronomy and Astrophysics*. 51 (1): 21–61.
[460] Sonjanovic, G. 2011. Probing the origin of neutrino mass: from GUT to LHC.
[461] Grossman, Y. 2003. TAST 2002 lectures on neutrinos.
[462] Koide, Y., Nuovo Cim. A 70 (1982) 411 [Erratum-ibid. A 73 (1983) 327].
[463] Sumino, Y. (2009). "Family Gauge Symmetry as an Origin of Koide's Mass Formula and Charged Lepton Spectrum". Journal of High Energy Physics. 2009 (5): 75. arXiv:0812.2103.
[464] Fire, A., Xu, S., Montgomery, M., Kostas, S., Driver, S., Mello, C. Potent and specific genetic interference by double-stranded RNA in Caenorhabditis elegans. Nature, 1998, 391 (6669): 806–811.
[465] Bernstein E., Caudy AA, Hammond SM, Hannon GJ (Jan 2001). Role for a bidentate ribonuclease in the initiation step of RNA interference. Nature. 409 (6818): 363–6.
[466] Valeri, N., et al. Epigenetics, miRNAs, and human cancer: a new chapter in human gene regulation. Mamm Genome (2009) 20:573-580.
[467] Iyer, L.M., Koonin, E.V., and Aravind L. Evolutionary connection between the catalytic subunits of DNA-dependent RNA polymerases and eukaryotic RNA-dependent RNA polymerases and the origin of RNA polymerases. BMC Struct Biol. 2003 Jan 28;3:1.
[468] Forterre, P. Displacement of cellular proteins by functional analogues from plasmids or viruses could explain puzzling phylogenies of many DNA informational proteins. Mol Microbiol. 1999 Aug;33(3):457-65.
[469] Legendrea, M., et al. In-depth study of Mollivirus sibericum, a new 30,000-yr old giant virus infecting Acanthamoeba. PNAS, September 8, 2015, E5327–E5335.
[470] Iyer, L.M., Balaji, S., Koonin, E.V., and L. Aravind. Evolutionary genomics of nucleo-cytoplasmic large DNA viruses. Virus Research 117 (2006) 156–184.

[471] Oliveira, G.P., et al. Promoter Motifs in NCLDVs: An Evolutionary Perspective. Viruses 2017, 9, 16.
[472] Kobiler, O., Drayman, N., Butin-Israeli, V. and A. Oppenheim. Virus strategies for passing the nuclear envelope barrier. Nucleus 3:6, 2012; 526–539.
[473] Fay, N. and N. Panté. Nuclear entry of DNA viruses. Frontiers in Microbiology 6(467), 2015.
[474] Cohen, S., Au, S., and N. Panté. How viruses access the nucleus. Biochimica et Biophysica Acta 1813 (2011) 1634–1645.
[475] Wang, G., Shimada, E., Zhang, J., Hong, J.S., Smith, G.M., Teitell, M.A, and C. M. Koehler. Correcting human mitochondrial mutations with targeted RNA import. PNAS 109(13), 2012, 4840–4845.
[476] Shaheen, H.H. and A. K. Hopper. Retrograde movement of tRNAs from the cytoplasm to the nucleus in Saccharomyces cerevisiae. PNAS 102(32), 2005, 11290–11295.
[477] O'Neill, R.E. Jaskunas. R., Blobeli, G., Palese, G.P., and J. Moroianu. Nuclear Import of Influenza Virus RNA Can Be Mediated by Viral Nucleoprotein and Transport Factors Required for Protein Import.The Journal of Biological Chemistry 270(39), 1995, 22701-22704.
[478] Marfori, M., et al. Molecular basis for specificity of nuclear import and prediction of nuclear localization. Biochimica et Biophysica Acta 1813 (2011) 1562–1577.
[479] Bryant, H.E., Wadd, S.E., Lamond, A.I., Silverstein, S.J., and J.B. Clements. Herpes Simplex Virus IE63 (ICP27) Protein Interacts with Spliceosome-Associated Protein 145 and Inhibits Splicing prior to the First Catalytic Step. J. of Virology 75(9), 2001, 4376-4385.
[480] Dubois J, Terrier O, Rosa-Calatrava M. Influenza viruses and mRNA splicing: doing more with less. mBio 5(3):e00070-14. 2014. doi:10.1128/mBio.00070-14.
[481] De Maio FA, Risso G, Iglesias NG, Shah P, Pozzi B, Gebhard LG, et al. The Dengue Virus NS5 Protein Intrudes in the Cellular Spliceosome and Modulates Splicing. PLoS Pathog 12(8): e1005841. 2016. doi:10.1371/journal.ppat.1005841
[482] Rogozin, I.B., Carmel, L., Csuros, M. and E.V. Koonin. Origin and evolution of spliceosomal introns. Biology Direct 2012, 7:11.
[483] Forterre, P. Genomics and early cellular evolution. The origin of the DNA world. C R Acad Sci III. 2001; 324(12):1067-76.
[484] Forterre, P. Three RNA cells for ribosomal lineages and three DNA viruses to replicate their genomes: A hypothesis for the origin of cellular domain. PNAS, 2006, 103(10), 3669-3674.

[485] Zong, J., Yao, X., Yin, J., Zhang, D., and H. Mac. Evolution of the RNA-dependent RNA polymerase (RdRP) genes: Duplications and possible losses before and after the divergence of major eukaryotic groups. Gene 447(1), 2009, 29-39.
[486] Sigova, A., Rhind, N., and P.D. Zamore. A single Argonaute protein mediates both transcriptional and posttranscriptional silencing in Schizosaccharomyces pombe. Genes & Dev. 2004. 18: 2359-2367.
[487] Takemura, M. Poxviruses and the origin of the eukaryotic nucleus. J Mol Evol. 2001 May; 52(5): 419-25.
[488] Bell P.J. Viral eukaryogenesis: was the ancestor of the nucleus a complex DNA virus? J Mol Evol. 2001 Sep, 53(3):251-6.
[489] Vellai T., Takacs K., and Vida G. A new aspect to the origin and evolution of eukaryotes. J Mol Evol. 1998 May;46(5):499-507.
[490] Gaudin, M., et al. Extracellular membrane vesicles harbouring viral genomes. Environ Microbiol. 2014 Apr;16(4):1167-75.
[491] Forterre, P. and M. Gaia. Giant viruses and the origin of modern eukaryotes. Curr Opin Microbiol. 2016 Jun; 31:44-9.
[492] Hale, C.R., Zhao, P., Olson, S., Duff, M.O., Graveley, B.R., Wells, L., Terns, R.M., and M.P. Terms. RNA-Guided RNA Cleavage by a CRISPR RNA-Cas Protein Complex. Cell, 2009 Nov, 139, 945–956.
[493] Makarova, Kira S, et al. Unification of Cas Protein families and a simple scenario for the origin of CRISPR/cas Systems. Biology Direct 2011:6.
[494] Kostyrka G. What roles for viruses in origin of life scenarios? Stud Hist Philos Biol Biomed Sci. 2016, Oct;59:135-44.
[495] Tessera, M. Origin of Evolution versus Origin of Life: A Shift of Paradigm. Int. J. Mol. Sci. 2011, 12, 3445-3458.
[496] Boyer M, et al., 2009. Giant Marseillevirus highlights the role of amoebae as a melting pot in emergence of chimeric microorganisms. Proc Natl Acad Sci U S A. 2009 Dec 22; 106(51):21848-53.
[497] Boyer, M., Madoui, M. A., Gimenez, G., La Scola, B., & D. Raoult. 2010. Phylogenetic and phyletic studies of informational genes in genomes highlight existence of a 4 domain of life including giant viruses. PLoS One. 2010 Dec 2;5(12):e15530.
[498] Nasir, A., Kim, K. M., & G. Caetano-Anolles. 2012. Giant viruses coexisted with the cellular ancestors and represent a distinct supergroup along with superkingdoms Archaea, Bacteria and Eukarya. BMC Evol. Biol. 12:156.
[499] Yutin, N., Wolf, Y. I., Raoult, D., & E. V. Koonin. 2009. Eukaryotic large nucleo-cytoplasmic DNA viruses: clusters of

orthologous genes and reconstruction of viral genome evolution. Virol. J. 6:223.
[500] Bell, P. J. 2001. Viral eukaryogenesis: was the ancestor of the nucleus a complex DNA virus? J. Mol. Evol. 53(3): 251-256.
[501] Iyer, L. M., Aravind, L., & E. V. Koonin. 2001. Common origin of four diverse families of large eukaryotic DNA viruses J. Virol. 75(23):11720-34.
[502] Saini, H. K., & D. Fischer. 2007. Structural and functional insights into Mimivirus ORFans BMC Genomics. 8: 115.
[503] Jagus, R., Bachvaroff, T. R., Joshi, B., & A. R. Place. 2012. Diversity of Eukaryotic Translational Initiation Factor eIF4E in Protists. Comp. Funct. Genomics. 2012:134839.
[504] Bell, P. Meiosis: Its Origin According to the Viral Eukaryogenesis Theory. http://dx.doi.org/10.5772/56876.
[505] Winters-Hilt, S. "Machine-Learning based sequence analysis, bioinformatics & nanopore transduction detection." ISBN: 978-1-257-64525-1. (2011).
[506] Feynman, R.P. There's plenty of room at the bottom: An invitation to enter a new field of physics. Engrg Sci 23:22 (1960).
[507] Winters-Hilt, S. "The Nanoscope." ISBN: 978-0-578-22307-0. (2019).
[508] D.B. Craig, E. Arriaga, J.C.Y. Wong, H. Lu, and N.J. Dovichi. The life and death of a single enzyme molecule. *Analytical Chemistry* 70, 39A-43A (1998).
[509] D.B. Craig, E. Arriaga, J.C.Y. Wong, H. Lu, and N.J. Dovichi. Studies on single alkaline phosphatase molecules: reaction rate and activation energy of a reaction catalyzed by a single molecule and the effect of thermal denaturation-The death of an enzyme. *Journal of the American Chemical Society* 118, 5245-5253 (1996).
[510] D.Y. Chen and N.J. Dovichi. Single-molecule detection in capillary electrophoresis: molecular shot noise as a fundamental limit to chemical analysis. *Analytical Chemistry* 68, 690-696 (1996).
[511] In-Geol Choi and Sung-Hou Kim. "Evolution of protein structural classes and protein sequence families". PNAS 103.38 (2006), pp. 14056-61. DOI: 10.1073/pnas.0606239103. PMID: 16959887.
[512] Bettina Wagner, Donald C. Miller, Teri L. Lear,and Douglas F. Antczak. The Complete Map of the Ig Heavy Chain Constant Gene Region Reveals Evidence for Seven IgG Isotypes and for IgD in the Horse. J Immunol 2004; 173:3230-3242.

[513] Daryl Fernandes. Demonstrating Comparability of Antibody Glycosylation during Biomanufacturing. European Biopharmaceutical Review. Summer 2005. pp 106 -110.
[514] Arstila, TP; Casrouge, A; Baron, V; Even, J; Kanellopoulos, J; Kourilsky, P. "A direct estimate of the human alphabeta T cell receptor diversity". Science. 286 (5441): 958-61. (1999).
[515] Winters-Hilt, S. “Immune Repertoire Profiling Using Nanopore Transduction,” Patent Pending 2019.
[516] Sakmann B, Neher E. Patch clamp techniques for studying ionic channels in excitable membranes. Annu Rev Physiol. 1984;46:455-72.
[517] "The Nobel Prize in Physiology or Medicine 1991". nobelprize.org. Nobel Media AB.
[518] Winters-Hilt, S. and A. Stoyanov. Nanopore Event-Transduction Signal Stabilization for Wide pH Range under Extreme Chaotrope Conditions. Molecules 2016, 21(3), 346 (2016)
[519] Winters-Hilt, S: Nanopore Detector based analysis of single-molecule conformational kinetics and binding interactions. BMC Bioinformatics 2006, Sept. 26, 7 S2: S21.
[520] Winters-Hilt, S. Nanopore Transducer Engineering and Design. Int J MolBiol Med 2(1): doi http://dx.doi.org/10.16966/ijmbm.108 (2017).
[521] Middlebrook R.D. (1989). "Null Double Injection and the Extra Element Theorem". *IEEE Transactions on Education*. 32 (3): 167–180. doi:10.1109/13.34149.
[522] Zhang, H. Reversal of HIV-1 Latency with Anti-microRNA Inhibitors. Int J Biochem Cell Biol. 2009 Mar; 41(3): 451–454.
[523] Liang, Ruqiang, David J Bates, and Eugenia Wang. Epigenetic Control of MicroRNA Expression and Aging. Curr Genomics. 2009 May; 10(3): 184–193.
[524] Bantounas, I., L A Phylactou1 and J B Uney. RNA interference and the use of small interfering RNA to study gene function in mammalian systems. J Mol Endocrinol. 2004 Dec;33(3):545-57.
[525] Agarwal, Vikram, George W Bell, Jin-Wu Nam, David P Bartel. Predicting effective microRNA target sites in mammalian mRNAs. eLife 2015;4:e05005.
[526] Yang W, Chendrimada TP, Wang Q, Higuchi M, Seeburg PH, Shiekhattar R, Nishikura K. Modulation of microRNA processing and expression through RNA editing by ADAR deaminases. Nat Struct Mol Biol (2006) 13: 13-21.
[527] Thomson K, Amin I, Morales E, and Winters-Hilt S. Preliminary Nanopore Cheminformatics Analysis of Aptamer-Target Binding Strength. BMC Bioinf. 8 S7, S14 (2007).

[528] Winters-Hilt S. The alpha-Hemolysin Nanopore Transduction Detector -- single-molecule binding studies and immunological screening of antibodies and aptamers. BMC Bioinf. 8 S7, S12 (2007).
[529] Leibniz, Gottfried Wilhelm. *Monadology.*
[530] Tipler, F. J. (1994). "The Physics of Immortality: Modern Cosmology, God and the Resurrection of the Dead." *Doubleday*..
[531] Miller, J. (2018). "The Omega Point: A Philosophical and Scientific Exploration." *Journal of Consciousness Studies*, 25(1), 1-20.
[532] Ricoeur, P. (1981). "Hermeneutics and the Human Sciences.".
[533] Teilhard de Chardin, P. (1955). "The Phenomenon of Man.".